CAMBRIDGE PHYSICAL SERIES

# THE THEORY

## OF

# EXPERIMENTAL ELECTRICITY

# CAMBRIDGE
## UNIVERSITY PRESS

University Printing House, Cambridge CB2 8BS, United Kingdom

Published in the United States of America by Cambridge University Press, New York

Cambridge University Press is part of the University of Cambridge.

It furthers the University's mission by disseminating knowledge in the pursuit of education, learning and research at the highest international levels of excellence.

www.cambridge.org
Information on this title: www.cambridge.org/9781107658455

© Cambridge University Press 1923

First edition 1905
Second edition 1912
Third edition 1923
First published 1923
First paperback edition 2014

A catalogue record for this publication is available from the British Library

ISBN 978-1-107-65845-5 Paperback

# THE THEORY

OF

# EXPERIMENTAL ELECTRICITY

BY

## WILLIAM CECIL DAMPIER WHETHAM,

M.A., F.R.S.

FELLOW AND SOMETIME SENIOR TUTOR OF TRINITY COLLEGE, CAMBRIDGE

*THIRD EDITION*

CAMBRIDGE

AT THE UNIVERSITY PRESS

1923

**ERRATUM**

P. 283, *for* Lane *read* Laue.

# PREFACE TO THE FIRST EDITION

FOR the past few years the writer of this book has given a course of lectures on the elementary theory of experimental electricity to a class of students at Trinity College, Cambridge. Experience, together with an occasional study of lecture note-books, has indicated that, to supplement the lectures, some definite and permanent statement is required—some book of reference, to which the students may turn for further elucidation of points not clear to them.

Thus the alternatives arose either of adapting the lectures to the lines of treatment of an existing book, or of writing a book which should correspond with the stage now reached in the evolution of teaching, which has extended over some ten years.

The great shift in the chief points of interest of experimental electricity, due to recent development in physical science, has changed the proportion of the various branches of the subject, and has put out of date many of the older standard text-books. To the phenomena of electrolysis, of conduction through gases, and of radio-activity, the physicist will now turn for knowledge newly acquired, for knowledge in the making, and for unsurveyed territory ready and waiting for the explorer. The writer has a firm belief in the advantage of giving to University students, even to those who confine themselves to an elementary study of their subjects, some insight into methods of research, together with some idea of recent results, and of unsolved problems ripe for examination. In this way interest is aroused, and natural science is presented in its true light, as a living and growing branch of knowledge—imperfect indeed and incomplete as all human creations must be, but more fascinating by reason of its very faults and limitations, and the consequent possibility of unexpected and beneficent development.

In the opinion of the writer, education in physical science may proceed profitably on two lines only—either by keeping in close touch with the advance of pure knowledge or by instruction which brings into prominence the application of science to technical industry. Unless one or other of these courses be followed, the study of natural science by adult students is in danger of becoming dead and unprofitable both to teacher and learner. If the teacher be in touch both with the practical problems which arise in the technical application of his science and with the investigations into the unknown which are extending the bounds of pure knowledge, so much the better for himself and for his pupils; but, at least, let him try to keep in sight of one of them. Let him train his students as though all were to become engineers or investigators: those who end by becoming teachers will be all the better for the misunderstanding. The state of a science in which the instruction is only fit to train each succeeding generation to become the teacher of the next recalls the economic condition of the famous island where the inhabitants lived by taking in each other's washing.

Of those who approach physics from the experimental side, a large majority study simultaneously the kindred science or chemistry. From their point of view, it is an advantage that the branches of the subject of electricity now developing most rapidly are those in which the connexion with chemistry is most intimate. Here, again, the comparatively small space allotted to electrolysis, conduction through gases, and radio-activity in the usual text-books is unsatisfactory. The unity of nature is best impressed on the observer by leading him to see the connexion between the different aspects in which, for the sake of convenience, our mental model of the world is presented. The fact that certain phenomena meet the student in his chemical work, is an argument, not for omitting their consideration from a course on physics, but for studying them with exceptional care, and for tracing their physical relations in the light of chemical knowledge.

In the following pages no attempt has been made to present a complete treatment of the science of electricity. The book is meant to be suggestive rather than exhaustive, to be an impressionist sketch rather than a finished picture. It aims at bringing into prominence those features which strike the writer as essential, without wearying the reader with a mass of unnecessary detail. An educational work is better too short than too long; better when in some points leaving curiosity unsatisfied than when attaining an ill-digested completeness. The object of the present undertaking has been to implant a thorough and clear knowledge of those physical principles necessary for an appreciation of the newer parts of the subject. All digressions, though interesting perhaps to the mathematician or experimenter, have been curtailed. The book is meant as an organic structure, each part of which has a definite and inevitable relation with the whole, each section its bearing on the plot of the story. To some extent, even a scientific text-book perforce must be a piece of literature and a work of art. Whether that necessity be welcomed or not, nothing is lost by keeping it clearly in mind.

The writer wishes to thank his wife, whose help, as always, has been freely given. Mr Norman Campbell, Fellow of Trinity College, has read the proof-sheets of the book. Mr G. F. C. Searle, of Peterhouse, has given his advice on several points; while the influence and inspiration of Professor J. J. Thomson will be evident throughout the work. In places the treatment is based on the writer's treatise on the *Theory of Solution,* and in others on his book on *The Recent Development of Physical Science.* Acknowledgment of this fact is due to the publishers of those two works, the Cambridge University Press, and Mr John Murray respectively.

W. C. D. W.

Trinity College, Cambridge,
*May* 15th, 1905.

# PREFACE TO THE SECOND EDITION

In putting forth a second edition of this book, the writer wishes to acknowledge the kindness of Mr G. F. C. Searle, whose suggestions have led to a modification of the text in several places.

Some additions have been made to the later chapters, where the advance in knowledge, for which the past six years have been responsible, has come within the scope of the work.

W. C. D. W.

Trinity College, Cambridge,
*November 2nd*, 1911.

# PREFACE TO THE THIRD EDITION

The preparation of the third edition has involved somewhat extensive additions and alterations. My thanks are specially due to the Cambridge and Paul Instrument Company, who have given most valuable advice about modern apparatus and lent many of the diagrams, and to Mr C. D. Ellis, my successor in the Lectureship at Trinity College, who has suggested improvements, especially in the later chapters.

W. C. D. W.

*November 26th*, 1922.

# CONTENTS

## CHAPTER I

## CHAPTER II

## CHAPTER III

## CHAPTER IV

## CHAPTER V

# CHAPTER VI

# CHAPTER VII

# CHAPTER VIII

# CHAPTER IX

# CHAPTER X

# CHAPTER XI

# CHAPTER XII

# CHAPTER I

## GENERAL PRINCIPLES OF ELECTROSTATICS

Early History. Franklin's work. Electric induction. Fluid theories of electricity. Variation of electric forces with distance. Quantity of electricity. Electric force or electric intensity. Electric potential. Electric capacity. Distribution of electricity on the surface of conductors. Specific inductive capacity or dielectric constant.

**1.** It was known to many in ancient Greece and Rome that amber and some other substances possessed, when rubbed, the power of attracting light bodies. Moreover, the shocks received on touching the fish called the torpedo were described by Pliny, and had been used to cure gout; though it is safe to surmise that no connexion was suspected between these apparently unrelated phenomena, or between either of them and the thunderbolts of Jove. Lucretius, who commented on the latter phenomena, hesitated to ascribe them to Divine interposition, observing that temples, even of Jove, far from being exempt, were especially liable to the visitation.

Though references to these facts, and speculations as to their nature, are scattered throughout the writings of the Middle Ages, no real advance on the knowledge of the ancients was made till Dr Gilbert (1540—1603) repeated the experiments with a view to finding some explanation.

William Gilbert, a native of Colchester, Fellow of St John's College, Cambridge, and sometime President of the College of Physicians, was one of the earliest and most distinguished of our English men of science—a man whose work Galileo himself thought enviably great. He was appointed Court physician to Queen

Elizabeth and James I., and a pension was settled on him to set him free to continue his researches in Physics and Chemistry.

It is to Gilbert that we owe the name electricity, which he derived from the Greek word ἤλεκτρον, amber. By investigating the forces on a light metallic needle, balanced on a point, he extended the list of electric bodies, and found also that many substances, including metals and natural magnets, showed no attractive forces when rubbed. He noticed that dry weather with north or east wind was the most favourable atmospheric condition for exhibiting electric phenomena—an observation liable to misconception till the difference between conductor and insulator was understood.

Gilbert's work was followed up by Robert Boyle (1627—1691), the famous natural philosopher who was once described as "father of Chemistry, and uncle of the Earl of Cork." Boyle was one of the founders of the Royal Society when it met privately in Oxford, and became a member of the Council after the Society was incorporated by Charles II. in 1663. He worked frequently at the new science of electricity, and added several substances to Gilbert's list of electrics. He left a detailed account of his researches under the title of *Experiments on the Origin of Electricity.*

The first to note that light and sound accompanied strong electric excitation was Otto von Guericke (1602—1686) of Magdeburg, who mounted a sulphur ball on a revolving axis and rubbed it with the hand. With this primitive electric machine he repeated the earlier experiments on a larger scale, and made the important discovery that, when once a light body had touched an electrified substance, it was thereupon repelled till it had touched some other object, when attraction again supervened.

In 1729 Stephen Gray discovered the electric properties of conductors and insulators, and, in conjunction with Wheeler, conveyed the electricity from a piece of rubbed glass over a distance of 886 feet, through a packthread suspended by silk loops. He found that the experiment failed if loops of hemp or wire were used, and also noted the conducting power of fluids and of the human body. Desaguliers added to these results the observation that conductors appeared not to be electrified by friction, a conclusion he would not have reached had he supported the conductors on insulating supports. Desaguliers' observation led to conductors being called non-electrics, a name they kept for many years.

About the same time Dufay, working in France, showed that two bodies electrified by contact with rubbed glass, or two bodies electrified from resin, repelled each other, while one of the first named would attract one of the latter. He was thus led to recognise two kinds of electricity—vitreous and resinous. Gray, who repeated his experiments, had already noted that the body used as rubber acquired as intense an electrification as did the object rubbed; and also, supported by some of his fellow-workers, suggested the identity of the "electric fire" with lightning.

The electric machine was now improved and increased in power, mainly by German and Dutch philosophers. A prime conductor, to collect the electric charge as formed, was added, and sparks of sufficient intensity to ignite spirits of wine were obtained.

The condenser originated in consequence of observations on the leakage of charge. This leakage, which always took place in open air and especially in damp air, gave rise to the notion that the charge might be preserved if surrounded by a non-conductor. The condenser seems to have been invented by more than one person.

From Muschenbroech of Leyden it took its name of the Leyden jar, and by him its properties were discovered in the attempt to electrify water in a glass bottle held in the hand. Dr Bevis suggested coating the outside of the jar with a metallic covering, and such a jar filled with water seems to have been used by Franklin and others. When both inside and outside had been coated with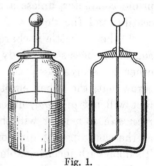

Fig. 1.

tin-foil, at the suggestion of Sir William Watson, the Leyden jar was complete. (Fig. 1.)

Watson directed a series of experiments for the Royal Society with a view to measuring the velocity of electricity. The velocity was of course much too great to be answerable to the experimental appliances of the time, and, even through more than 12,000 feet

of wire, propagation was pronounced to be instantaneous. With
reference to these experiments, Watson remarks "that when the
observers have been shocked at the end of two miles of wire,
we infer that the electrical circuit is four miles, viz. two miles
of wire and the space of two miles of the non-electric (conducting)
matter between the observers, whether it be water, earth, or
both." This seems to be the first occasion when the earth was
used and recognised as a return circuit.

2. Watson suggested that the two kinds of electrification
Franklin's    might represent the excess and defect in a single
work.         kind of electricity, and this "single fluid theory,"
with its positive and negative charges, was elaborated and made
more definite by Dr Benjamin Franklin of Philadelphia (1706—
1790), journalist, philosopher, and statesman. Franklin showed
that the two coatings of a Leyden jar were oppositely electrified,
and that, in the terms of the one-fluid theory, as much electricity
escaped from the outer coating as entered the inner coating. "The
phial will not suffer what is called a charging unless as much fire
can go out of it one way as is thrown in by another. A phial
cannot be charged standing on wax or glass, or hanging on the
prime conductor, unless a communication be form'd between its
coating and the floor.

"When a bottle is charged in the common way its *inside* and
*outside* surfaces stand ready, the one to give fire by the hook, the
other to receive it by the coating; the one is full and ready to
throw out, the other empty and extremely hungry; yet as the
first will not *give out*, unless the other can at the same instant
*receive in*, so neither will the latter receive in, unless the first can
at the same instant give out. When both can be done at once
'tis done with inconceivable quickness and violence."

The ideas prevalent at that time regarded the electricity stored
in the jar as analogous to a liquid stored in a bottle; but Franklin,
setting himself to determine the essential features of the jar,
found that similar effects could be obtained by placing two sheets
of lead, one on each side of a plane sheet of glass. By gilding
the glass in front of a picture of the King, Franklin was able,
by means of a violent electric shock, to frustrate the efforts of

" the conspirators " who endeavoured to take off a little moveable gilt crown from the King's head. In after years he may have remembered this playful experiment when, his efforts to prevent a quarrel having failed, Franklin, both in his own land and as ambassador to France, did much to remove from the King's crown the bright jewel of the American colonies.

Franklin found that a sharp point near an electrified body caused it to lose its charge, and noticed the glow which such a point presents in a darkened room. From experiments like these he reached the conception of an electric atmosphere surrounding charged bodies, and concluded that the charge resided on the surface. The Leyden jar, moreover, showed phenomena which indicated that the charge resided, not on the coatings, but in the glass itself.

" The whole force of the bottle, and power of giving a shock, is in the GLASS ITSELF; the non-electrics in contact with the two surfaces serving only to *give* and *receive* to and from the several parts of the glass; that is, to give on one side, and take away from the other.

" This was discovered here in the following manner. Purposing to analyse the electrified bottle, in order to find wherein its strength lay, we placed it on glass, and drew out the cork and wire which for that purpose had been loosely put in. Then taking the bottle in one hand, and bringing a finger of the other near its mouth, a strong spark came from the water, and the shock was as violent as if the wire had remained in it, which showed that the force did not lie in the wire. Then to find if it resided in the water, being crowded into and condensed in it, as confin'd by the glass, which had been our former opinion, we electrified the bottle again, and placing it on glass, drew out the wire and cork as before; then taking up the bottle we decanted all its water into an empty bottle, which likewise stood on glass; and taking up that other bottle, we expected if the force resided in the water to find a shock from it; but there was none. We judged then that it must either be lost in decanting or remain in the first bottle. The latter we found to be true: for that bottle on trial gave the shock, though filled up as it stood with fresh unelectrified water from a teapot. To find, then, whether glass had this property merely as glass, or whether the form contributed anything to it, we took a pane of sash glass and laying it on the stand placed a plate of lead on its upper surface ; then electrified that plate, and bringing a finger to it, there was a spark and shock.  We then took two plates of lead of equal dimensions, but less than the glass by two inches every way, and electrified the glass between them, by electrifying the uppermost lead ; then separated the glass from the lead, in doing which what little fire might be in the lead was taken out and the glass being touched in the electrified parts with a finger, afforded only very small pricking sparks, but a great number of them might be taken from different places.  Then dexterously placing it again between the leaden plates, and completing a circle between the two surfaces, a violent shock ensued.—Which demonstrated the power to reside in glass as glass, and that the non-electrics in contact served only, like the armature of a loadstone, to unite the force of the several parts and bring them at once to any point desired : it being a property of a non-electric that the whole body instantly receives or gives what electrical fire is given or taken from any one of its parts."

In the last sentence Franklin clearly shows that he understood the essential nature of the then-called non-electrics as conductors of electricity.  The experiment with the Leyden jar makes plain the paramount part played in electrical phenomena by the non-conductor, or, as it was then termed, the electric *per se.*

This experiment is of fundamental importance, for the modern theory of electricity, as we shall see, regards the dielectric as an essential seat of electrical manifestations.  On this foundation Faraday and Maxwell laid the corner-stones of electrical science.

As soon as the spark and noise of an electric discharge were noticed, their resemblance to lightning and thunder was recognised, and the identity in nature of the two phenomena suspected.  The problem of the establishment of this identity seems to have possessed a fascination for the mind of Franklin, and many of his later letters are filled with the description of experiments repeating on a small scale, with the charges of Leyden jars, the effects of lightning in fusing metals, rending materials, &c.  The discharging action of points suggested to Franklin the idea of the lightning-conductor.

"May not the knowledge of this power of points be of use to mankind, in preserving houses, churches, ships, &c. from the stroke of lightning, by directing us to fix on the highest parts of those edifices upright rods of iron made sharp as a needle, and gilt to prevent rusting, and from the foot of those rods a wire down the outside of the building into the ground, or down round one of the shrouds of a ship, and down her side till it reaches the water? Would not these pointed rods probably draw the electrical fire silently out of a cloud, before it came near enough to strike, and thereby secure us from that most sudden and terrible mischief?"

Franklin then goes on to suggest that in order "to determine the question whether the clouds that contain lightning are electrified or not" an iron rod should be erected on some high tower or steeple. When thunder-clouds passed, sparks might be drawn from the lower end of the rod.

The letters containing this suggestion led to its adoption in France, England, and other countries, with complete success— a success, too complete indeed in the case of Professor Riehmann of St Petersburg, who was killed by a shock from an iron rod erected on his house. Meanwhile Franklin himself had safely carried out a similar experiment by means of a kite.

"To the top of the upright stick of the kite is to be fixed a very sharp pointed wire, rising a foot or more above the wood. To the end of the twine, next the hand, is to be tied a silk ribbon, and where the silk and twine join a key may be fastened. This kite is to be raised when a thunder-gust appears to be coming on, and the person who holds the string must stand within a door or window, or under some cover, so that the silk ribbon may not be wet; and care must be taken that the twine does not touch the frame of the door or window. As soon as any of the thunder-clouds come over the kite the pointed wire will draw the electric fire from them, and the kite, with all the twine, will be electrified, and the loose filaments of the twine will stand out every way and be attracted by an approaching finger. And when the rain has wet the kite and twine, so that it can conduct the electric fire freely, you will find it stream out plentifully from the key on the approach of your knuckle. At this key the phial may be charged; and from electric fire thus obtained spirits may be kindled, and all the other electric experiments be performed, which are usually done by the help of a rubbed glass globe or tube, and thereby the sameness of the electric matter with that of lightning completely demonstrated.'

During the eighteenth century, many experiments were made on the electrification produced by heating certain minerals and crystals, such as tourmaline; and attention was drawn once more to the benumbing power of shocks given by the torpedo and certain other fish. Their electrical organs were examined, and the shocks they inflict were ascribed definitely to electrical manifestations.

**3.** While Franklin pursued his researches in America, an English philosopher and schoolmaster, John Canton

Electric induction.

(1718—1772), born at Stroud in Gloucestershire, was working on parallel lines in London. In 1750 he received a medal from the Royal Society in recognition of his improvements in artificial magnets, and in 1752 he was the first in England to verify his friend Franklin's demonstration of the identity of lightning and electricity.

But perhaps the greatest of Canton's discoveries was the process of electrification by induction, which he explained in terms of the theory of electrical atmospheres. If an insulated conductor, let us say of a cylindrical form, is placed in the neighbourhood of a charged body, the nearer end of the cylinder acquires a charge of the opposite sign, and the farther end a charge of the same sign as that on the body. The charges may be demonstrated by fitting the cylinder with vertical metal stands, from which hang linen threads carrying pith balls. An electric charge repels the balls from the stands (Fig. 2). If the cylinder, still insu-

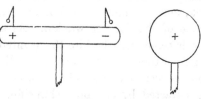

Fig. 2.

lated, be removed from the neighbourhood of the original charge, the charges on the cylinder neutralize each other. If, however, while under the influence of the electrified body, the cylinder be touched with the finger or otherwise connected with the earth, the charge of the same name as that of the inducing body escapes, and the cylinder retains the charge of the opposite name. As long as the electrified body is near, this induced charge remains fixed or bound : it will not escape even along a conductor. When, however, the electrified body and the cylinder are separated, the induced charge becomes free ; the cylinder acts as an ordinary charged system—it will share its charge with any other insulated conductor in contact with it, and lose its charge to earth if a passage be opened. These phenomena admit of an obvious explanation in terms of the attraction between unlike charges and the repulsion between those of the same name, the original uncharged cylinder being imagined to contain equal and opposite quantities of " vitreous and resinous " or of " positive and negative " electricity.

Electrostatic induction gives us the best means of obtaining a continuous series of electric charges, and on this principle all modern electric machines are based.

The simplest influence machine is the electrophorus of Volta, invented in 1775. It consists of a plate of resin or vulcanite which is electrified by rubbing it with cats' fur. A brass plate, held on an insulating handle, is then placed upon it, and touched with the finger. The negatively electrified vulcanite induces a positive charge on the lower surface of the brass plate and repels the corresponding negative charge to earth. The finger is then

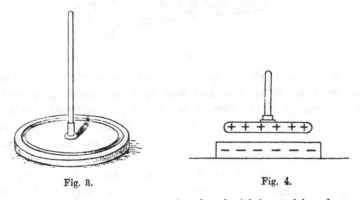

Fig. 3.                              Fig. 4.

removed, leaving the brass plate insulated with its positive charge. If it be now raised from the cake of vulcanite, work is done against the electric forces, the positive charge is set free, and can be communicated to other conductors. The energy of the electric charge obviously is derived from the work of electric separation performed in raising the charged plate from the oppositely electrified vulcanite; as long as the two charges are in close proximity, they are unavailable, and possess no energy. The charge on the vulcanite is not sensibly affected by the working of the instrument, since the brass plate touches it at a few points only. Thus the brass plate can be replaced, and the process repeated till the original charge is lost by leakage. A continuous series of charges is obtained at the expense of manual labour.

Fig. 5.

Similar principles are applied in the powerful influence machine invented by Wimshurst and now used extensively in physical laboratories (Fig. 5). Two shellac-coated glass plates are made to revolve in opposite directions by means of a hand-wheel or motor. Sectors of thin brass or tin-foil are fixed to the outer surfaces of the plates as indicated in the figure. These sectors are touched as they revolve by wire brushes on the ends of two uninsulated diagonal conductors. To start the machine, a charged ebonite rod is held opposite one of these brushes. The negative charge on the rod, acting through the two glass plates, induces a positive charge on the sector, the corresponding negative charge escaping through the brush. The positively charged sector passes on, till it comes opposite the brush on the other side of the machine. Here it induces a negative charge on one of the sectors of the other plate then in contact with the brush, and finally gives up its charge to one of the collecting combs which are connected with the terminals of the machine. Thus, each charged sector acts as an inducing agent for sectors on the opposite plate before rendering up its charge. Once started, the machine increases its own charges, till leakage or discharge between the terminals prevents further accumulation. Large Wimshurst machines, with several pairs of plates, are now often worked by electric motors, and give a constant stream of opposite charges of very great intensity.

4. As we have seen, the phenomena of electricity known in the middle of the eighteenth century were explained by the single fluid theory of Watson and Franklin. An alternative explanation was, however, suggested by other physicists. According to this second view, the two kinds of electrification were to be explained as the manifestation of two imponderable fluids, co-existent but with opposite properties, which neutralized each other. But the age was not ready for such fundamental and detailed theories of electricity ; in fact, even now we are not in a position to accept finally any theory as to its ultimate nature, though it is certainly now more probable that electricity is to be thought of as particles rather than as a fluid. Nevertheless, the single and double fluid theories were most useful working models, and played the true part of scientific hypotheses in enabling observers to describe and coordinate the phenomena, and to find new and profitable fields of research.

At the period we are now considering, the knowledge already acquired might be summarized by saying that, in unelectrified bodies, equal quantities of the opposite electric fluids existed, and that, by the friction of certain substances, a separation between the fluids might be effected. The fluids passed easily through the interstices of conductors or non-electrics, substances which, when held in the hand, could not be electrified by friction, but only moved with great difficulty through electrics or non-conductors. Each electric fluid tended to spread over the surface of any

*Fluid theories of electricity.*

conductor ; the different portions of the fluid seemingly exerted a repulsive force on each other.  Moreover, two bodies similarly electrified repelled each other.  On the other hand, opposite electricities on the same conductor tended to approach and neutralize each other, while two bodies oppositely electrified were subject to a force of attraction.  Electric charges at rest were found only on the surface of conductors, but in motion they used the whole substance of the conductor and not merely its surface.  Those doubtful on this point might, according to Franklin, be convinced by the passage of an electric shock through the substance of their bodies.

The properties of condensers or Leyden jars were explained by the attraction of the opposite charges collected or condensed on their coatings ; though, as we have seen, Franklin had already shown that the glass itself played the more essential part in the process.

As soon as we attempt to deal with the fluids of the two-fluid theory as other than mathematical abstractions, the difficulty of the conception becomes manifest.  We have to suppose that the mixture of two fluids in equal proportions gives us something so devoid of properties that it cannot be detected.  The one-fluid theory, as developed by Franklin, avoids this difficulty.  In his view, portions of the single positive fluid attract ordinary matter and repel other portions of electricity.  Unelectrified matter is supposed to be associated with so much of the electric fluid that the attraction of external electricity for the matter is just balanced by the repulsion for the normal charge associated with it.  Excess of the fluid beyond the normal charge means positive electrification, defect means negative electrification.  There is much in Franklin's theory which simplifies the explanation of electrical phenomena, though even now we do not know the true relation between positive and negative electrification.

**5.**  At this stage of the science, the most important phenomena were the attractions and repulsions between electrified bodies, and the investigation of these forces became the immediate object of experimenters.  The Newtonian law of gravitation must have suggested the possibility that,

Variation of electric forces with distance.

in electric phenomena also, we are dealing with forces which vary inversely as the square of the distance. It was shown by Æpinus, a German philosopher who lived from 1724 to 1806, that the force diminished as the distance increased, but a satisfactory demonstration of the truth of the inverse square law was first published by Robison and soon after by Charles Augustin Coulomb of Montpellier (1736—1806), a distinguished military engineer in the service of France, and a member of the Académie des Sciences. Coulomb's researches were carried on by means of an instrument of his own invention, known as the torsion balance. Nevertheless, as we shall see presently, a more accurate method than that of Coulomb had been worked out and used previously by Cavendish, though with characteristic reticence he had not thought fit to give his discovery to the world.

The original idea of measuring small forces by balancing them against the torsion of a wire seems to have been due to the Rev. John Michell, whose method was applied by Cavendish in 1797 to determine the gravitational attraction between balls of lead. Coulomb, probably unaware of Michell's suggestion, investigated experimentally the couple required to twist a wire through a given angle, and found that it varied as the angle and as the fourth power of the diameter of the wire. He then constructed a balance, such as is illustrated in Fig. 6, and used it to investigate electric forces.

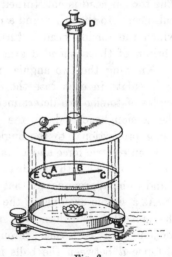

Fig. 6.

To the centre of the top of a cylindrical glass case a vertical glass tube is fixed. At the upper end of the tube is fitted a graduated torsion head from which hangs a fine silver wire. The wire carries a thin straw, covered with sealing wax, with a small pith ball at one end and a vertical disc of oiled paper (not shown in the figure) at the other. The disc acts as a damper of the oscillations of the bar, and as a pointer by means of which the position of the bar is ascertained on a graduated circle surrounding the middle of the glass case. A second hole, towards the side of the top of the case, admits a vertical rod of shellac, suspended by a clip and carrying a second conducting ball, which is thus placed in a fixed position within the case, such that the centre of the ball lies between the zero of the graduations on the case and the lower end of the hanging wire. The fixed ball being removed, the torsion head is adjusted till the suspended ball lies with its centre at the same point. The fixed ball is then reinserted, and, when the moveable ball has come to rest in contact with it, both are electrified by means of a small metal ball carried on an insulating handle. The moveable ball is repelled, and takes up a position of equilibrium under the electric repulsion and the force of torsion. Its position on the graduated circle is then noted. The torsion head is now turned round so as to force the moveable ball nearer to the other, and a note is made of the angle through which the torsion head is turned and the new position of equilibrium of the suspended arm.

Knowing the two angular positions of equilibrium, it is easy to calculate in each case the distances between the balls. The angles of torsion are determined by observing the graduations on the torsion head and on the glass case. The torsional couple being proportional to the angle of twist in the wire, a relation between electric force and distance is obtained. If the balls are small compared with the distance between them, the relation is found to be approximately that of the inverse square.

As a simple example of the method, let us imagine that, after the moveable ball has been repelled through an angle $\alpha$, the torsion head is turned through an angle $\beta$ till the original distance $d$ between the balls is halved. The angular deviation of the moveable arm is now less than $\alpha$, let us suppose $\gamma$, and the total twist in the wire is $\gamma + \beta$ instead of $\alpha$, as at first. It will now be found that $\gamma + \beta$ is approximately four times $\alpha$, showing that halving the distance has increased the force in the ratio of 1 to 4. In a similar

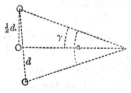

Fig. 7.

manner, diminishing the distance to one-third will be found to increase the torsional and electric forces in the ratio of 1 to 9.

While such experiments serve to show that the law of inverse squares is not far from the truth, they are not susceptible of any great accuracy. The law, as we shall see later, is only exact if the dimensions of the charged bodies are infinitely small compared with the distance between them, or if the bodies consist of spheres uniformly electrified. Now, although the bodies in Coulomb's apparatus are spheres, they are not uniformly electrified, for the repulsive action between the charges concentrates them on the farthest parts of the spheres. Besides this difficulty, the unavoidable leakage of charge during the course of the measurements renders the experiment uncertain in result. The torsion balance, in less experienced hands than those of Coulomb, is very far from being an instrument of precision.

Fortunately we are not dependent on the torsion balance for an exact verification of the law of inverse squares, and, as already stated, another method had been devised and carried out for his own satisfaction by Henry Cavendish some years before the publication of Coulomb's results. Cavendish discovered many things which are now associated with the names of other men, and it was only when his unpublished manuscripts, belonging to the Duke of Devonshire, were edited by Clerk Maxwell in 1879, that the world understood to what an extent he had anticipated the conclusions of more recent times. As Maxwell says, "Cavendish cared more for investigation than publication. He would undertake the most laborious researches in order to clear up a difficulty which no one but himself could appreciate, or was even aware of, and we cannot doubt that the result of his enquiries, when successful, gave him a certain degree of satisfaction. But it did not excite in him that desire to communicate the discovery to others which, in the case of ordinary men of science, generally ensures the publication of their results. How completely these researches of Cavendish remained unknown......is shown by the external history of electricity." In the present age of publicity we may perhaps overrate the eccentricity of Cavendish's character, but, even in his own more leisurely days, that eccentricity seems to have arrested attention.

Cavendish's method of examining the law of force depends on

the experimental proof that there is no electric force within a charged sphere. Newton in his *Principia* had demonstrated that, on the assumption of an inverse square law, the gravitational force would vanish within a uniform spherical shell of gravitating matter. It is easy to transfer the proof to the case of uniform electrification, which is mathematically similar.

Let us consider the resultant force at the point $P$ within the sphere. With $P$ as common apex, draw two small equal-angled cones with their axes in the same straight line on opposite sides of $P$. The bases of these two small cones being very small, we may take the area $A$ of one base to be at a constant distance from $P$, and to act as an electrified particle. By hypothesis, therefore, its charge will act at $P$ with a force proportional to $1/(PA)^2$; similarly the charge

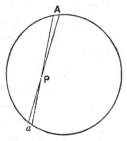

Fig. 8.

on $a$ will act with a force proportional to $1/(Pa)^2$. But, the sphere being uniformly electrified, the charges on these areas $A$ and $a$ will be proportional to the areas. Now, the bases being equally inclined to the axes of the cones, they will possess areas directly proportional to the squares $(PA)^2$ and $(Pa)^2$. Thus, the forces at $P$ due to the charges on $A$ and $a$ will increase with distance as much as they diminish, and will be the same as though the charges were at the same distance from $P$, when obviously the forces they exert are equal and opposite. The force at $P$ due to the two elementary cones, then, vanishes; and by dividing the whole surface of the sphere into elementary areas by similar pairs of cones, it is clear that the force due to each pair of cones must vanish also, so that there is no resultant force.

We have now shown that the assumption of the law of inverse squares leads to the disappearance of electric force within a uniformly charged sphere. It remains to prove that no other law is consistent with this result. Through a point $P$, inside a sphere with centre $O$ (Fig. 9), draw a plane at right angles to $OP$. If the charge on the part of the sphere above this plane produces an

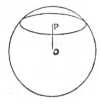

Fig. 9.

attraction in the direction $OP$, that below the plane will act in the direction $PO$. When the law is that of the inverse square, these two forces balance each other. If the force decrease with increase of distance at some rate higher than the square, the force due to the electrification of the larger area below the plane will fall off at a rate too fast to be compensated by the greater effect of the larger area. The resultant force will therefore act in the direction $OP$ towards the smaller area. On the other hand, if the force diminish less fast than the square of the distance increases, the greater distance from $P$ of the larger area will be overcome by the effect of the larger charge, and there will be a resultant force in the reverse direction.

It follows, therefore, that no power of the distance other than the inverse square is consistent with the condition of no electric force within the uniformly charged sphere. Maxwell pointed out that the experimental fact, which was known to Cavendish, that the distribution of electricity was similar on similar figures, irrespective of size, shows that the law of force must involve some power of the distance and no other mathematical function. If, then, it can be shown experimentally that there is no electric force within a uniformly charged sphere, it must follow that the inverse square is the only possible law of force. This result was first given in a general form by Laplace, though, as we have stated, owing to the known experimental properties of similar bodies, Cavendish's assumption of some power of the distance does not sacrifice the generality of his proof.

Whenever an electric force acts on a conductor, as we have seen above, electric separation occurs, and parts of the conductor become electrified differently to other parts. If we find that, within a charged conductor, no separation of electricity occurs, it shows that no electric force exists.

To examine this point, Cavendish says, "I took a globe 12·1 inches in diameter, and suspended it by a solid stick of glass run through the middle of it as an axis, and covered with sealing-wax to make it a more perfect non-conductor of electricity[1]. I then inclosed this globe between two hollow pasteboard

[1] Glass is hygroscopic, and in damp weather becomes covered with a conducting film of moisture.

hemispheres, 13·3 inches in diameter...in such manner that there could hardly be less than $\frac{4}{10}$ of an inch distance between the globe and the inner surface of the hemispheres in any part, the two hemispheres being applied to each other so as to form a complete sphere, and the edges made to fit as close as possible, notches being cut in each of them so as to form holes for the stick of glass to pass through.

"By this means I had an inner globe included within a hollow globe in such a manner that there was no communication by which the electricity could pass from one to the other.

" I then made a communication between them by a piece of wire run through one of the hemispheres and touching the inner globe, a piece of silk string being fastened to the end of the wire by which I could draw it out at pleasure.

"Having done this I electrified the hemispheres by means of a wire communicating with the positive side of a Leyden phial, and then, having withdrawn this wire, immediately drew out the wire which made a communication between the inner globe and the outer one, which, as it was drawn away by a silk string, could not discharge the electricity either of the globe or hemispheres. I then instantly separated the two hemispheres, taking care in doing it that they should not touch the inner globe, and applied a pair of small pith balls, suspended by fine linen threads, to the inner globe, to see whether it was at all over or undercharged[1].

"The result was that though the experiment was repeated several times I could never perceive the pith balls to separate or show any signs of electricity......

"Hence it follows that the electric attraction and repulsion must be inversely as the square of the distance, and that when a globe is positively electrified the redundant fluid in it is lodged entirely on its surface......

"In order to form some estimate how much the law of the electric attraction and repulsion may differ from that of the inverse duplicate ratio of the distances without its having been perceived in this experiment" Cavendish tested the sensitiveness of his apparatus by communicating to the inner sphere an amount of electricity which was just appreciable with the pith balls, and was

---

[1] *i.e.* positively or negatively electrified.

equal to a known fraction of that communicated to the outer sphere in the experiment described. "We may conclude that the electric attraction and repulsion must be inversely as some power of the distance between that of the $2 + \frac{1}{50}$th and that of the $2 - \frac{1}{50}$th, and there is no reason to think that it differs at all from the inverse duplicate ratio."

A similar experiment was carried out, using a hollow rectangular box instead of the folding hemispheres, with a similar result. Whatever the form of the conductor, as long as it is completely closed, there is no electric force within it.

Truly this research is no less remarkable as a model of scientific method, than for the importance of the results obtained.

The pair of pith balls used by Cavendish do not constitute a sensitive electroscope. A more delicate instrument, invented by Bennet, consists of two gold-leaves suspended from a wire placed within a glass jar. When electrified, the leaves repel each other and diverge at an angle which roughly indicates the intensity of electrification. In recent years the gold-leaf electroscope has become an instrument of accurate research, and is employed for

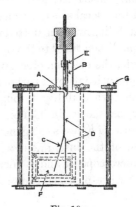

Fig. 10.                                    Fig. 10 a.

measuring the electrical conductivity of air and other gases under the influence of Röntgen rays, radio-active substances, etc. In one form, due to Mr C. T. R. Wilson, a single strip of gold-leaf $C$ is attached to a brass plate $D$ (Fig. 10 a), from which the leaf is repelled when electrified.

Another even more sensitive electrometer, devised by Mr Wilson and Dr Kaye, consists of an insulated gold-leaf $L$ (see Fig. 11) which, hanging from $C$, is attracted out of the vertical position by a plate $P$ attached to the case of the instrument. The case, and $P$ with it, can be tilted and adjusted to any angle by a screw. It will

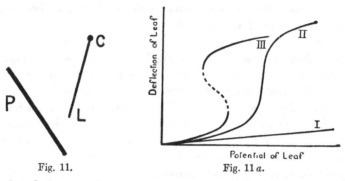

Fig. 11.                    Fig. 11 $a$.

be found that, as the angle is varied, the equilibrium of the leaf can be made either stable over the whole of its path of deflexion or unstable over part of it. Figure 11 $a$ shows three curves illustrating the possible cases. Curve I shows complete stability and

small sensitivity, and Curve III high local sensitivity with instability over the dotted region. When tilted to the right angle, the electrometer gives high sensitivity over part of its path with stability throughout, as shown in Curve II. Figure 11 $b$ shows the general view of the instrument with a microscope $M$ and a micro-

Fig. 11 $b$.

meter eye-piece, on the scale of which the position of the gold-leaf is read.

In the "string electrometer," originally suggested by Professor Einthoven, the moving system is a silvered quartz fibre, tightly stretched between and parallel to two oppositely electrified metal plates. The minute force, about $10^{-8}$ dyne, needed to deflect the quartz fibre is of the same order as that acting on the gold-leaf of a sensitive electroscope. The capacity is very small, the motion of the fibre is dead-beat and the period short. The instrument is therefore suitable for following rapid changes in voltage.

With a gold-leaf electroscope, it has been possible to obtain a much more exact confirmation of the law of inverse square than was effected by Cavendish. Faraday, for instance, constructed a wooden cube covered with tin-foil, large enough to contain himself and his electroscopic instruments. The cube was supported on insulating feet, and intensely electrified. Even when brush and spark discharges were darting from the outside, no electrification could be detected within.

Using Lord Kelvin's quadrant electrometer (§ 20) Maxwell, about the year 1870, repeated Cavendish's experiment, and came to the conclusion that the power of the distance involved in the law of force cannot differ from 2 by more than the 1/21600th part.

The fact that the charge on a system resides on the outside only may be demonstrated roughly in many ways, one of which is a modification of Cavendish's experiment with the separable hemispheres (Fig. 12).

Besides its chief interest as a means of verifying the law of inverse square, the fact that there is no electric force inside a closed conductor of any form, carrying any distribu-

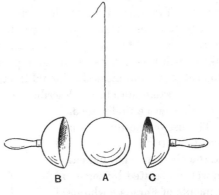

B     A     C

Fig. 12.

tion of electricity at rest, shows that such a conductor serves to screen points within it from electrostatic disturbances due to outside electrification. For effective screening, it is not necessary to have a continuous conductor. A cover of wire-gauze, as shown in Fig. 10, or even strips of tin-foil pasted on a glass shade, are usually sufficient protection to electric instruments placed within. The indications of such instruments are not affected by charges without the screen.

**6.** In order to acquire the power of measuring definitely any
Quantity of physical quantity, it is necessary to find some satis-
electricity. factory method of defining a unit of the quantity,
and to show that, in certain conditions at all events, the quantity
we are dealing with does not arbitrarily increase or decrease—
that, in certain circumstances, it is a conservative quantity. Thus,
mass and energy are quantities. Units can be devised, and two
masses or two quantities of energy can be added together so that
the joint mass or energy is the sum of the parts. The temperature
of a mixture of two volumes of water, on the other hand, is not the
sum of those of the two before junction. In this sense, temperature
is not a physical quantity.

A knowledge of the laws of electric attraction and repulsion,
and of their variation with the distance, enables us to define a
convenient unit for quantity of electricity. In terms of this unit,
it is found that all known phenomena are consistent with the
supposition that electricity may be treated as a real physical
quantity. Two quantities of electricity of the same kind can be
given to the same insulated conductor, and, as measured by its
external electric forces, that conductor, allowing for the loss by
leakage, will then possess a charge represented by the sum of the
two charges communicated. Should the two charges be of opposite
kinds, the same statement describes the facts if by the sum the
algebraic sum be understood.

The use of the term "quantity of electricity" is justified
and the phenomena of induction
are well illustrated by some experiments of Faraday, who used
a pewter ice-pail to represent a
nearly closed conductor. The
ice-pail was placed on an insulating stand and connected with
an electroscope by means of a
wire. A brass ball, suspended by
a long thread of white silk, was
electrified, and gradually lowered
into the pail. As it approached,

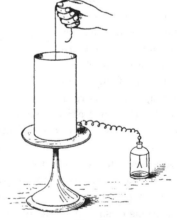

Fig. 13.

the leaves of the electroscope diverged, and the divergence increased till the ball was well inside the vessel, some few inches below the top. The divergence of the leaves then became constant, however the ball was moved about within the pail. If its charge were shared with another ball, no change appeared in the indications of the electroscope. Finally, if the ball touched the inside surface of the vessel, again the divergence of the gold-leaves was unaffected. The whole of the charge on the ball now had passed to the outside of the surrounding vessel, and this was verified by removing the ball, when the leaves maintained their deflection, and the ball was shown to be unelectrified.

Two chief conclusions may be drawn from these facts:

(1) The deflection of the electroscope depends on something which remains constant however the charged ball be moved, when that ball is once fairly within the vessel, and however the charge be distributed; we may call this something a quantity of electricity, the quantity remaining constant, and not depending on the nature, position, or size of the charged body.

(2) Let us then accept the idea of a quantity of electricity, and the fact that, on contact with the inside, the whole of the charge on the ball passes to the outside. It follows from the constancy of the deflection on contact that the quantity of electricity distributed on the inside of the vessel at first, owing to induction, must be equal and opposite to the charge on the ball, since, when they coalesce, they leave the charge on the outside unaffected. Touching the inside of the vessel with the ball only leaves the deflection unaltered if the charge be effectively surrounded on all sides by the vessel. Thus, for the induced charge to be equal as well as opposite to the inducing charge, the outside conductor must completely surround the charged body.

In the light of our knowledge of the laws of electric force, the electrostatic unit of quantity of electricity may be defined conveniently as that quantity which, when placed on a small particle at unit distance from a similarly electrified particle, repels it with unit force when the two particles are separated by air. In the system of units usually adopted, the unit of distance is the centimetre, and the unit of force the dyne. Now, it can be shown experimentally, by means of the torsion balance for instance, that two such units in conjunction produce a doubled force on the repelled unit—the electric forces vary as the product of the quantities of electricity and inversely as the square of the distance between them. Writing $e_1$ and $e_2$ for the two quantities of electricity, and $r$ for the distance, we have

$$F = \frac{e_1 e_2}{r^2} \quad \dots\dots\dots\dots\dots\dots\dots\dots(1),$$

for the force between them when separated by air. We shall see presently that, in the general case, when any insulating medium lies between the charges, another factor must enter into this expression.

7.    In the equation given above, $F$ denotes the mechanical force acting between two similarly electrified particles. It **Electric force or electric intensity.** must not be confused with another quantity, of different physical dimensions, which is called the electric force or the electric intensity. Let us suppose that, in some way or other, an electric field is set up. An electrified body placed in that field is subject to forces, but the amount of force it suffers depends on the charge on the body. It is convenient to have some quantity to express the state of the field independently of any charge used to explore it. The electric intensity at any point of the field is defined as the mechanical force per unit of charge exerted on a particle charged with an infinitesimal quantity of positive electricity and placed at the point. If we denote the electric intensity by $f$, the mechanical force on a body carrying a charge $e$ is

$$F = fe,$$

so that the dimensions of $f$ are those of mechanical force divided by those of quantity of electricity.

Electric intensity, like mechanical force, is a vector quantity—it involves the idea of direction. A line drawn through an electric field so as always to coincide in direction with the electric intensity is known as a line of force. In another place we shall have much to say about lines of force.

8.    The early experimenters soon found that it was impossible **Electric potential.** to electrify a system without limit. As the process continued, leakage increased, and eventually luminous discharges appeared, which carried away the charge as fast as it

was imparted to the system. With a given source of electricity, this stage was reached sooner with small systems than with large ones. Hence arose the ideas of quantity of electricity, of the electric capacity of the receiving system, and of the degree of electrificatión, or what we now term the potential.

Cavendish seems to have been distinguished from his predecessors and contemporaries by a clear grasp of the meaning of the phrase "degree of electrification" and by an exact definition of it. In terms of his fluid theory, he boldly gave it a physical meaning as the pressure of the fluid. The relative degrees of electrification determine in which direction electricity will pass when two bodies are connected by means of a conductor, and, when equilibrium has been attained, the potential of the bodies must be the same: all parts of a conducting system must be at the same potential when its charge is at rest. In this sense electric potential is analogous to temperature in heat, or to pressure in hydrodynamics.

The difference of electric potential between two points may be defined as the work done against the electric forces in bringing one positive unit of electricity from one point to the other, the movement of the unit charge being supposed to produce no appreciable change in the potential. The electric potential of any single point in space then is best defined as the amount of work it is necessary to do against the electric forces in bringing up to the point a particle charged with one unit of positive electricity from a point at an infinite distance from all charges, where, by arbitrary definition, we may suppose the potential to be zero.

As the simplest and most useful case, let us calculate the electric potential due to an isolated positive charge $e$ at a distance $r$ from it.

Let $O$ denote the position of the point-charge $e$. At a point $P$

O                     P Q R S

Fig. 14.

the electric intensity, *i.e.* the force on a unit positive charge, is $e/(OP)^2$ and at a neighbouring point $Q$ it is $e/(OQ)^2$. If $P$ and $Q$ are taken very near together no sensible error will be made if the force throughout the distance $PQ$ is taken as constant and as equal to the geometrical mean $e/(OP \cdot OQ)$.

The work done in bringing a positive unit from $Q$ to $P$ is then

$$PQ \frac{e}{OP \cdot OQ} = (OQ - OP) \frac{e}{OP \cdot OQ} = e\left(\frac{1}{OP} - \frac{1}{OQ}\right).$$

Similarly, the work from $R$ to $Q$ is $e\left(\dfrac{1}{OQ} - \dfrac{1}{OR}\right)$, and from $S$ to $R$, $e\left(\dfrac{1}{OR} - \dfrac{1}{OS}\right)$.

Now, the total work done from $S$ to $P$ is the sum of the elements of work given above, and is therefore $e\left(\dfrac{1}{OP} - \dfrac{1}{OS}\right)$.

This process could be continued from as far as we please. If we imagine it begun at an infinite distance, the expression for the work done throughout, or the electric potential, becomes

$$V = e\left(\frac{1}{OP} - \frac{1}{\infty}\right) = \frac{e}{OP} = \frac{e}{r}.$$

In the notation of the integral calculus, this proof may be written

$$V = \int_{\infty}^{r} \frac{e}{r^2}\, dr = \frac{e}{r} - \frac{e}{\infty} = \frac{e}{r} \quad\ldots\ldots\ldots\ldots\ldots(2).$$

Thus the potential due to a point-charge is proportional to that charge, and, since any distribution of electrification may be supposed to be made up by a collection of point-charges, it follows that, if all the charges in a system vary together, the potentials and the charges are always proportional to each other.

If the electric intensity be constant over a certain length $l$, then the work done in carrying a positive unit over the distance $l$ is $fl$, and this is the difference of potential $V$. That difference is proportional to $l$, and thus, if the force be constant, the rate $V/l$ of change of potential per unit length in the direction of the force is constant also, and equal to the electric intensity.

If the electric intensity be not constant, its value at any point will be equal to the maximum rate of change of the potential at that point; that is, to the difference of potential over a small distance in the direction of the intensity divided by that distance, when the distance taken is so small that the force is sensibly constant over it.

The potential diminishes as we pass away from a positively electrified body, and the force on a positive unit is also directed away from such a body. Thus the resultant electric intensity acts in the direction in which the potential decreases. In general terms, the electric intensity in any direction is equal to the rate of fall of potential in that direction.

In terms of the differential calculus

$$f_x = -dV/dx \quad \dots\dots\dots\dots\dots\dots\dots(3).$$

**9.** The charge on an isolated conductor is proportional to its
Electric capacity. potential, and the constant ratio between them, or the quantity of electricity needed to raise the potential by unity, is defined as the electric capacity of the conductor. Here the system is really made up of the conductor and the far-off surrounding conductors, which are taken to be at zero potential. We are, in fact, concerned, not with an isolated body, but with a system made up of the body, the dielectric field surrounding it, and the conducting boundaries of that field with an equal and opposite charge residing on them. The capacity $C$ of this system is then given by

$$C = \frac{e}{V_1 - V_2},$$

where $e$ is the charge on one boundary of the field, and $V_1 - V_2$ denotes the difference of potential between the boundaries. A similar expression gives the definition of the capacity for systems such as condensers, which are essentially the same as the case just considered.

Cavendish again appears as the first to possess a clear idea of this quantity, and to make definite measurements of the capacity of different bodies. He constructed condensers by pasting tin-foil on each side of glass sheets, and arranging them in sets of three, so that one of the second set had the same capacity as three of the first set, and so on. As we shall see hereafter, the capacity of an isolated conducting sphere placed in air is numerically equal to its radius, and Cavendish used the diameter of a sphere possessing the same amount of charge as the given body, when at the same degree of electrification, as the measure of its capacity. Thus when he says that a certain body contains $n$ "inches of electricity" we may interpret his result as meaning that its capacity is $\frac{1}{2}n$ inch.

The electric capacity of an insulated conductor is not a property of the form and dimensions of that conductor alone. It depends on the neighbouring conductors, and, as will be shown later on, on the nature of the dielectric or insulator separating it from the neighbouring conductors. The effect of one conductor on the capacity of another can be demonstrated easily by means of a gold-leaf electroscope. The divergence of the leaves measures the degree of electrification, *i.e.* the difference of potential between the gold-leaf system and its surroundings. If an uninsulated brass plate be held in the hand, and be moved nearer to the plate at the top of the electroscope, the divergence of the leaves decreases, showing that their potential is diminished. No electric communication with the gold-leaf system has been made; hence the charge is unaltered and the capacity of the upper plate must have been increased.

10.   The laws of force and induction being given, the calcula-
Distribution of    tion of the distribution of charge on the surface of a
electricity on
the surface of   conductor of any form is a question of mathematics.
conductors.
        For an isolated sphere, everything being sym-
metrical, the distribution is clearly uniform, the charge, as we have seen, residing on the outside of the sphere. As we pass over the surface of a conductor, places of great curvature will be found to be the most highly electrified. On sharp points, the surface density of electrification becomes very great; hence the power possessed by points of discharging a conductor.

The cases of an isolated ellipsoid, and of two spheres mutually influencing each other, were solved by Poisson, but it was reserved for Green to develop a more general method by means of which the distribution of charge on many other figures could be determined.

The next step was made by Lord Kelvin, who invented a new method—the theory of electrical images. When an otherwise isolated point-charge is placed in front of a plane or spherical conducting surface connected with the earth, Lord Kelvin showed how to find the distribution of induced charge on the surface. The resultant electric intensity at points in the space in front of the plane conductor is the same as though the induced charge on the plane were replaced by a point-charge placed behind the plane in the position of the optical image of the original point-charge.

While such investigations are of great interest in the mathematical theory of electricity, they do not bear directly on the side of the subject considered in this book. With this brief reference we shall pass on.

**11.** Cavendish observed that coated plates of glass contained more charge than would be expected from their thickness and area when compared with similar

Specific inductive capacity or dielectric constant.

systems of two metal plates separated by air. In certain cases, he gave numerical values for the ratio between the observed and the calculated charges, glass giving a value of about 8, and shellac about 4½. This phenomenon was rediscovered and investigated carefully by Faraday. Faraday used the apparatus shown in Fig. 15, consisting of two concentric spheres, the space between which could be filled with air or any other fluid. Solid dielectrics such as shellac and glass were examined in the form of hemispheres, which could be placed in the lower half of the space between the spheres of the apparatus.

Fig. 15.

Two arrangements of this kind were made, exactly similar to each other. One was charged by connecting the inner sphere with a Leyden jar, and the degree of electrification tested by means of a torsion balance. The charge was then shared with the other apparatus, and the common potential again tested with the torsion balance. If the capacity of the two pieces of apparatus be the same, the potential of the doubled capacity will be half that of the single one with the same charge. This was found to be the case when both were filled with air, but, when a hemisphere of glass or shellac was interposed, the final potential was less than half the original potential, showing that the final capacity was more than twice the capacity of the single apparatus filled with air. Allowing for the fact that only about half the inter-spherical interval was filled with the solid dielectric, Faraday calculated what the capacity would have been if the space had been filled completely. The ratio of this capacity to that of the apparatus when it contained air was defined as the specific inductive capacity of the solid dielectric. In these experiments, Faraday discovered, with certain solids, a slow creep of electric charge into the substance of the dielectric. The result of measurements of the capacity of condensers thus depends on the time during which the condenser remains charged, and differences exist between dielectric constants as determined by long-charge methods, like that of Faraday, and short-charge methods such as will be described later (§ 69).

*Dielectric Constants. Specific Inductive Capacities*[1].
*Long-Charge Values.*

*The value for air is taken as unity.*

| | | | | |
|---|---|---|---|---|
| Glass, light flint- | 6·57 | | Petroleum oil | 2·10 |
| „ very dense flint- | 10·1 | | Ozokerite | 2·13 |
| „ hard crown- | 6·96 | | Turpentine | 2·23 |
| „ plate- | 8·45 | | Benzene | 2·38 |
| Sulphur, non-crystalline | 3·84 | | Carbon bisulphide | 2·67 |
| Mica | 6·64 | | Ether | 4·75 |
| Ebonite | 3·15 | | Distilled water | 75 to 80 |
| Resin | 2·55 | | | |

Faraday could detect no difference between the specific inductive capacity of air at ordinary pressure and that of air at the lowest pressure he could reach. Electric forces act across the best vacuum we can obtain; exact experiments show that, taking the dielectric capacity of a vacuum as unity, that of air at atmospheric pressure is about 1·00059. For practical purposes, then, the value of the constant for air still may be considered to be unity.

It is possible to show, by direct experiment, that the force exerted by an electrified body on, for instance, a pith ball, is diminished by the interposition of a solid dielectric such as a sheet of

---

[1] Taken from Prof. Sir J. J. Thomson's *Recent Researches in Electricity*, p. 468.

paraffin ; and, as we shall see later, the effect observed in measuring capacities is explained if we write the force between the two point-charges placed in a medium extending to a great distance on all sides as

$$F = \frac{e_1 e_2}{k r^2},$$

where $k$ is Faraday's quantity, the specific inductive capacity or the dielectric constant. This, then, is the general expression for the law of force between electric charges. It is obvious that the corresponding value for the electric potential at a point due to a charge at a distance $r$ must, in its general form, be written as

$$V = \frac{e}{k r}.$$

## REFERENCES.

*Experiments and Observations on Electricity, made at Philadelphia in America*;
    by Dr Benjamin Franklin.
*The Works of Henry Cavendish*; edited by J. Clerk Maxwell.
*Experimental Researches in Electricity*; by Michael Faraday.

# CHAPTER II

## SOME THEOREMS OF ELECTROSTATICS

Total normal induction. Gauss' theorem. Electric intensity outside a uniformly charged sphere. Electric intensity outside a uniformly charged infinite plane. The electric capacity of an isolated sphere. Capacity of two concentric spheres. Capacity of two parallel planes. The mechanical force on a charged conductor. Energy of electrified systems. The quadrant electrometer.

**12.** In deducing the mathematical theory of electrostatics
Total normal from the results of experiment as formulated in the
induction. laws of force, we shall find the following theorem,
Gauss'
Theorem. due to Gauss, of great assistance.

Let us draw an imaginary closed surface, surrounding any quantity of electricity distributed in any manner. Then let us divide that surface into an immense number of very small elements of area, one of which is $a$, and calculate the value of the electric intensity (or the electric force) $N$ normal to each element of area. Assuming that the dielectric constant of the medium is $k$, we shall show that the value of the sum of all the areas multiplied by their corresponding normal forces and by the dielectric constant, or $\Sigma aNk$, is equal to $4\pi$ times the total quantity of electricity $e$ enclosed within the surface. We may state this result in the form

$$\Sigma aNk = 4\pi e.$$

The quantity $\Sigma aNk$ is called the total normal induction through the surface.

To simplify the equations, let us first take the medium to be air, for which $k$ is unity.

Let $PQ$ be the trace of one of the elements of area on the plane of the paper, and, at first, let all the charge be concentrated at the point $O$. With centre $O$ and distance $OP$ describe a sphere cutting $OQ$ in $R$.

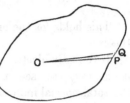

The element of area $PQ$ being very small, the electric intensity over it is uniform, acts along $PQ$, and is equal to $\dfrac{e}{OP^2}$. The component of this intensity

Fig. 16.

normal to the area is $\dfrac{e}{OP^2}\cos QPR$, since the angle $QPR$ is equal to the angle between $OP$ and the normal to the surface $PQ$.

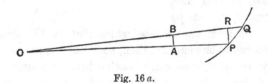

Fig. 16 a.

Thus the product of the area and the normal force is

$$\text{Area } PQ \times \frac{e}{OP^2}\cos QPR = \text{Area } PR \times \frac{e}{OP^2}.$$

With centre $O$ describe any sphere, cutting $OP$ and $OQ$ in $A$ and $B$.

Then            Area $AB$ : Area $PR$ :: $OA^2$ : $OP^2$

or            $$\frac{\text{Area } AB}{OA^2} = \frac{\text{Area } PR}{OP^2}.$$

Hence            $$\text{Area } PR \times \frac{e}{OP^2} = \text{Area } AB \times \frac{e}{OA^2}.$$

Therefore the product of the element of area and its normal force is the same for the original imaginary surface as it is for this sphere round the point-charge as centre.

The same proof holds for all elements of area; and thus $\Sigma aN$ is the same for the surface as it is for the sphere.

For the sphere, the normal electric force is constant, and is everywhere equal to the electric intensity.

Hence        $\Sigma aN = N \cdot \Sigma a = N \cdot 4\pi (OA)^2$

$$= \frac{e}{(OA)^2} \cdot 4\pi (OA)^2 = 4\pi e.$$

This holds for the original surface also, and we establish Gauss'
theorem,                    $\Sigma aN = 4\pi e.$

If the only electric charge in the field lies outside the surface,
it is easy to see that
the total normal induction
must vanish.

For each of the areas
$pq$ and $PQ$, the product
$aN$ is the same as for a
sphere drawn round $O$ as
centre, and, in this case,
the electric force at $pq$

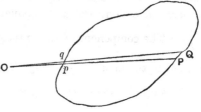

Fig. 16 *b*.

acts outwards from the surface while that at $PQ$ acts inwards.
With reference to the surface, then, they have opposite signs, and,
since they are equal, must cancel each other. In a similar manner,
the electric forces due to other pairs of elements of area cancel,
and thus the total normal induction due to the charge outside the
surface must vanish also.

If the charge inside or outside the surface be distributed,
instead of being concentrated at a point, it may be shown that
similar results hold good. Any electric distribution can be represented as made up of a number of point-charges $e_1$, $e_2$, $e_3$... etc.
with electric intensities, normal to one single element of area,
equal to $N_1$, $N_2$, $N_3$... etc. The sum of these components is
equal to $N$, the resultant component of the electric intensity
normal to the area. Thus

$\Sigma aN = \Sigma (N_1 + N_2 + N_3 + ...) a$

$\qquad = \Sigma N_1 a + \Sigma N_2 a + \Sigma N_3 a + ...$

$\qquad = 4\pi e_1 + 4\pi e_2 + 4\pi e_3 + ... = 4\pi (e_1 + e_2 + e_3 + ...) = 4\pi e.$

Therefore the total normal induction is equal to $4\pi$ times the total
charge within the surface however that charge be distributed.

If, instead of air, we have a uniform insulating medium with
dielectric constant $k$, the result becomes

$$\Sigma aN = \frac{4\pi e}{k}, \quad \text{or} \quad \Sigma aNk = 4\pi e.$$

**13.** Let us find the electric intensity at a point $P$ outside a uniformly electrified sphere, of which $O$ is the centre. With centre $O$ and distance $OP$ describe an imaginary spherical surface, and to this surface apply Gauss' theorem,

*Electric force outside a uniformly charged sphere.*

$$\Sigma \alpha N = 4\pi e.$$

The surface being spherical, and the charged sphere within it being uniformly electrified, everything is symmetrical, and the normal electric force must everywhere be equal to the resultant electric force. Thus $\Sigma \alpha N = f \Sigma \alpha$, where $f$ is the electric intensity at the point $P$.

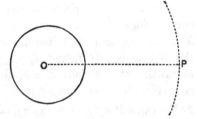

Fig. 17.

Hence   $f . 4\pi (OP)^2 = 4\pi e$

or   $$f = \frac{e}{(OP)^2}.$$

It follows, therefore, that outside a uniformly electrified sphere the electric force is equal to that which the same quantity of electricity would produce if concentrated at the centre of the sphere.

We have supposed the sphere to be surrounded by air. In the general case, if it be immersed in a medium with a dielectric constant $k$, the intensity at the point $P$, without it, is seen to be

$$f = \frac{e}{k\,(OP)^2}.$$

Let us call the charge per unit area on the surface of the sphere the surface density of electrification, and denote it by the symbol $\sigma$.

The electric intensity just outside a sphere of radius $r$, *i.e.* indefinitely near its surface, is

$$f = \frac{e}{kr^2} = \frac{4\pi r^2 \sigma}{kr^2} = \frac{4\pi \sigma}{k}.$$

For a point indefinitely near it, the form of the surface does not matter, and this result therefore gives the electric intensity

3—2

just outside a closed charged surface of any form. It is known as Coulomb's law, and can be deduced directly from Gauss' theorem.

**14.** Let $AB$ denote a part of a uniformly charged plane of infinite area. Outside the plane, with its axis

Electric inten-<br>sity outside<br>a uniformly<br>charged<br>infinite plane. normal to the plane, describe the cylinder $PQ$. From symmetry, the intensity must everywhere be normal to the plane, and uniform in any plane

parallel to the charged one. There is thus no normal electric force over the curved sides of the cylinder, and the force is uniform and normal over the flat ends. If $f_1$ and $f_2$ be the intensities at the ends $P$ and $Q$ respectively, by Gauss' theorem,

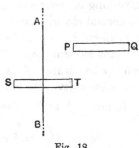

Fig. 18.

$\Sigma aN = (\text{area } P \times f_1) + (\text{area } Q \times f_2) = 0.$

Thus $f_1 = f_2$, since area $P = $ area $Q$. The intensity is therefore constant at all distances from the plane.

Describe another cylinder $TS$, with axis normal to the plane, enclosing an area $\alpha$ of the plane, on which the total charge is $\sigma\alpha$. By Gauss' theorem

$$(\text{Area } T \times f_1) + (\text{Area } S \times f_2) = \frac{4\pi\sigma\alpha}{k}$$

But $\qquad \text{Area } T = \text{Area } S = \alpha ;$

thus $$2\alpha f = \frac{4\pi\sigma\alpha}{k}$$

or $$f = \frac{2\pi\sigma}{k}$$

It should be noted here that $\sigma$ is the total amount of electricity on both sides of the plane per unit area. If $\sigma$ be taken as the charge on one side of the plane only, then, for an isolated plane,

$$f = \frac{4\pi\sigma}{k}.$$

In this book we shall adopt the former convention, and use $\sigma$ to denote the charge on unit area, counting in the charge on both sides of the plane.

**15.** We defined the capacity $C$ of an isolated conductor, or

The electric capacity of an isolated sphere. rather the capacity of the system formed of that conductor and of the far-off boundaries of the dielectric medium, as the charge $e$ required to increase the potential $V$ by unity. Hence

$$C = \frac{e}{V}.$$

The potential $V$ is the difference of potential between the conductor and the far-off boundaries of the medium, which by definition are at the zero of potential.

Since an isolated sphere is itself symmetrical and is unaffected by the action of other charged bodies, it must be electrified uniformly, and therefore will act at points outside as though its charge were concentrated at its centre. Thus the potential (§ 8) at its surface will be

$$V = \frac{e}{kr},$$

where $r$ is its radius, and $k$ the specific inductive capacity of the surrounding medium.

Hence $$C = \frac{e}{e/kr} = kr.$$

Thus, in air, where $k$ is taken as unity, the capacity of an isolated sphere is numerically equal to its radius.

**16.** The capacity of a condenser formed of two concentric

Capacity of two concentric spheres. spheres, with radii $r_1$ and $r_2$, and a dielectric with constant $k$ between them, is a quantity of practical as well as of historical interest. Let us suppose that a charge $+e$ be given to the inner sphere, and that the outer sphere is connected with the earth. A charge $-e$ will then be induced on it, since it surrounds the inner sphere completely. With regard to points outside both of them, each sphere acts as though its charge were concentrated at its centre. But the spheres have the same centre, and the charges are equal and opposite. Thus the potential just outside the outer sphere, and therefore the potential of its surface, must be zero.

The surface of the inner sphere, being a conductor, must be at uniform potential, for otherwise the charge on it could not be at rest. Since it is a closed surface which contains no charge, there is no electric force inside, and the potential is everywhere the same within it. We may therefore find the potential of the inner sphere by calculating the value at the centre. Here the charge $+e$ on the surface of the inner sphere is at a uniform distance of $r_1$, and the potential due to it is consequently $\dfrac{e}{kr_1}$. The charge $-e$ on the outer sphere is at a uniform distance of $r_2$, so that the total potential of the inner sphere is

$$V_1 = \frac{e}{kr_1} - \frac{e}{kr_2} = \frac{e}{k}\left(\frac{1}{r_1} - \frac{1}{r_2}\right).$$

The capacity of the double system is, by what was said in § 9, $e/(V_1 - V_2)$, and we have seen that $V_2$ the potential of the outer sphere is zero. Thus

$$C = \frac{e}{V_1 - V_2} = \frac{e}{\dfrac{e}{k}\left(\dfrac{1}{r_1} - \dfrac{1}{r_2}\right)} = \frac{r_1 r_2 k}{r_2 - r_1}.$$

**17.** Let $AB$ and $CD$ represent portions of two parallel plates of infinite extent. Let the area of each portion

Capacity of two parallel plates.

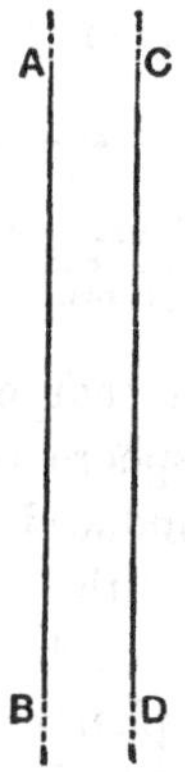

considered be $a$. Let one plate be insulated and charged with a quantity of positive electricity till the surface density is $\sigma$. If the other plate be connected with earth, a negative charge will be induced on it. If we imagine the plates to be small portions of two concentric infinite spheres, we see that the charge on the part of the sphere $CD$ which lies opposite the area $AB$ will acquire a charge numerically equal to that on the area $AB$. Thus the surface-density on $CD$ will be $-\sigma$. Owing to the attractive forces between the opposite charges, the whole electrification on each plate will reside on the side nearest to the other plate.

Each plate will produce an electric intensity outside it equal to $2\pi\sigma/k$ (§ 14). A positive unit placed between them will be repelled from one and attracted towards the other, so that the forces are in the same direction, and the total electric intensity between the plates is $4\pi\sigma/k$. Since we consider parts of infinite plates, everything is symmetrical, and the intensity is uniform. Thus the

Fig. 19.

difference of electric potential $V_1 - V_2 = ft = 4\pi\sigma t/k$, where $t$ is
the thickness of the dielectric stratum between the plates.

Now the capacity for an area $a$ is given by

$$C = \frac{e}{V_1 - V_2} = \frac{\sigma a k}{4\pi\sigma t} = \frac{a k}{4\pi t}.$$

This result is of great practical importance. As it stands,
it gives an approximate value for the capacity of all condensers,
in which a thin layer of dielectric lies between two conducting
surfaces. If the dielectric be thin, the form of the surfaces is
immaterial. Thus the capacities of Leyden jars, as well as those
of plate condensers, may be calculated approximately from this
formula.

Such arrangements, however, do not give us portions of infinite
planes, and, if accurate results are needed, a correction must be
applied to allow for the effects of the free edges of the system.
This correction requires complicated mathematical treatment. In
certain pieces of apparatus, however, it is possible to use a device
due to Lord Kelvin, which makes the edge-correction unnecessary.

The insulated plane, let us suppose in the form of a circular
plate, is surrounded by an annular
disc, or guard ring, lying in the same
plane, and separated only by a narrow
air-gap from the central plate. This
compound plate is then placed opposite
and parallel to another circular plate
with a diameter equal to that of the
guard ring, thus forming a double
plate condenser. The central plate
and its guard ring can be connected
by means of a fine wire, and then form

Fig. 20.

parts of the same conductor. In this state they are charged, and,
since the irregularities due to the edges are confined to the
guard ring, the electric distribution on the central plate $ab$ is
uniform. The wire is then removed, so that the central plate $ab$
is insulated. We can thus deal with the charge on the central
plate alone, and this charge is that calculated by the elementary
theory of capacity given above. The use of a guard ring enables
us to employ what are, in essence, portions of infinite planes.

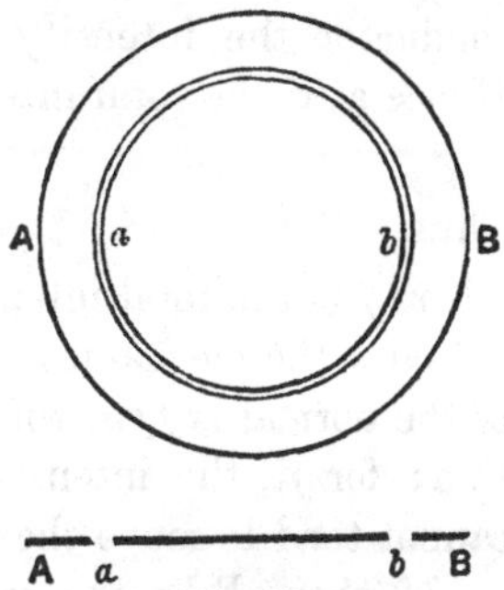

**18.** At a point indefinitely near the surface of a conductor, the neighbouring surface appears as a large plane; hence the electric intensity due to the charge on the neighbouring surface is normal to the surface, so that the mechanical force on a small area of the electrified surface will act along the normal.

*The mechanical force on a charged conductor.*

The electric intensity $f$ at a point $P$ in the small area may be considered to be composed of a part $f_1$ due to the charge on the small area, and of a part $f_2$ due to the charges on the rest of the conductor and on other conductors in the field. Thus

$$f = f_1 + f_2.$$

As we pass from $Q$ to $S$, points just without and just within the surface respectively, we change the sign of $f_1$ the intensity due to the charge on the small area round $P$. Neither the sign nor the value of $f_2$ will change, for, with respect to the more distant parts of the conductor, and to other conductors, $Q$ and $S$ are practically coincident points. Inside a closed charged conductor the intensity must vanish (§ 5). Hence at $S$ the total intensity

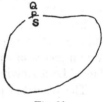

Fig. 21.

$$f_2 - f_1 = 0.$$

Thus $$f_1 = f_2 = \tfrac{1}{2} f,$$

where $f$ is the total intensity outside the surface at $Q$.

Now the mechanical force on the small area $\alpha$ in the direction of the normal is $f_2\alpha\sigma$, where $\sigma$ is the surface density of electrification; for $f_1$, the intensity due to the charge on the area itself, cannot tend to move the area.

If, then, $F$ be the mechanical force per unit area along the normal,

$$F\alpha = f_2\alpha\sigma = \tfrac{1}{2} f\alpha\sigma,$$

and $$F = \tfrac{1}{2} f\sigma.$$

Since (§ 13) by Coulomb's law $f = 4\pi\sigma/k$,

$$F = \frac{2\pi\sigma^2}{k} = \frac{f^2 k}{8\pi}.$$

In the case of two parallel plates separated by a dielectric layer, the mechanical force on the central plate ($ab$, Fig. 20) surrounded by a guard ring is of practical importance. The electric intensity due to the other opposite plate is $2\pi\sigma/k$, if the dielectric constant of the separating medium is $k$ (§ 14). The charge on the plate $ab$ produces no force on itself normal to its plane, thus $2\pi\sigma/k$ is the force on a positive unit of charge placed on the plate $ab$. The actual charge is $\sigma\alpha$, and the mechanical force is $2\pi\alpha\sigma^2/k$.

It is more useful to express this force in terms of the difference of potential $V_1 - V_2$ between the plates. The total intensity $f$ between the plates, due to the charges on both, is $4\pi\sigma/k$, and the intensity is uniform, so that $V_1 - V_2 = ft$, where $t$ is the distance between them. We may therefore write

$$V_1 - V_2 = 4\pi\sigma t/k$$

or
$$\sigma = \frac{(V_1 - V_2)k}{4\pi t}.$$

The mechanical force on the area $\alpha$ of $ab$ is

$$F = \frac{2\pi\alpha\sigma^2}{k} = 2\pi\alpha\frac{(V_1 - V_2)^2 k^2}{16\pi^2 t^2 k} = \frac{\alpha k (V_1 - V_2)^2}{8\pi t^2}.$$

It is interesting to compare this equation with the expression (§ 11) for the mechanical force between two charges, viz. $e_1 e_2/kr^2$. In any system, when the *charges* are constant, the forces are inversely proportional to $k$, and the substitution for air of a dielectric of high specific inductive capacity diminishes the forces. On the other hand when the *potentials* are kept constant, an increase of dielectric constant increases the mechanical forces between the parts of the system.

The expression for the force between two parallel planes is used in Lord Kelvin's Absolute or Trap-door Electrometer. In one form of this instrument, the moveable disc $S$ is attached to three light springs, of which two are shown in Fig. 22. The disc

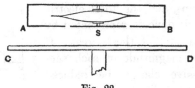

Fig. 22.

usually lies rather above the plane of the guard ring $AB$. The plate

$CD$ is screwed up, till the attractive force pulls $S$ into the plane of $AB$, as shown by a fixed sighting lens and a cross wire attached to $S$. The force then has a certain value, known from previous experiments on the weight needed to bring $S$ to its sighted position. Hence the difference of potential between $S$ and $CD$ is determined:

$$V_1 - V_2 = t\sqrt{\frac{8\pi F}{\alpha}},$$

the dielectric being air.

**19.** If no loss of energy occur, the energy of an electrified system must be equal to the work done against the electric forces during the process of charging. By the definition of electric potential, we know that, when unit positive charge is brought up to a conductor at a potential $V$, the work done against the electric forces is $V$, provided no appreciable change is thereby produced in the potential. Similarly if $e$ units be brought up, the work done is $eV$. If $e$ be small enough, the supposition of constant potential will be justified. Now we can imagine the process of charging an isolated conductor to be carried out by the successive addition of very small charges, and thus the electric work done may be written

*Energy of electrified systems.*

$$W = e_1 V_1 + e_2 V_2 + e_3 V_3 + \ldots\ldots\ldots$$

Let us represent this process graphically as in Fig. 23, where the ordinates represent potential and the abscissæ charge. The successive terms in the series just given are represented in the figure by the areas of the successive vertical strips, and the total electric work done is represented by the sum of these areas. If the magnitude of each successive charge be reduced without limit, this sum approximates to the area of the figure $OPM$.

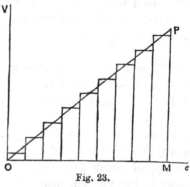

Fig. 23.

Now by § 8 it follows that the potential of an isolated conductor is proportional to the charge. Thus the curve $OP$ is a straight line, and the figure $OPM$ is a triangle. Its area is therefore $\frac{1}{2}eV$, where $e$ is the final charge, and $V$ the final potential. The electric work done in charging, and the electric potential energy stored in the conductor by virtue of its charge, are each given by

$$W = \tfrac{1}{2}eV.$$

An important case is that of the two parallel plates. The electrical energy is

$$W = \tfrac{1}{2}e(V_1 - V_2),$$

since this is the electrical work done in carrying a charge $e$ from one plate to the other, producing thereby the difference of potential.

We can express this energy in terms of the dimensions of the apparatus, and either the potentials or the charges:

(i)  Since the intensity $f$ between the plates is uniform,

$$V_1 - V_2 = ft,$$

where $t$ is the distance. Now $f$ is $4\pi\sigma/k$, if $k$ be the dielectric constant of the medium between the plates. Thus $\sigma$ is $\dfrac{fk}{4\pi}$, and

$$W = \tfrac{1}{2}\sigma\alpha(V_1 - V_2)$$

$$= \frac{1}{2}\frac{fk}{4\pi}\alpha(V_1 - V_2)$$

$$= \frac{V_1 - V_2}{t}\frac{k}{8\pi}\alpha(V_1 - V_2)$$

$$= \frac{\alpha k}{8\pi t}(V_1 - V_2)^2 \quad\dots\dots\dots\dots\dots(1).$$

(ii)  Again, we may express the energy in terms of the charge on one plate.

$$W = \tfrac{1}{2}e(V_1 - V_2)$$

$$= \tfrac{1}{2}e.ft = \tfrac{1}{2}e\frac{4\pi\sigma}{k}t$$

$$= \tfrac{1}{2}e\frac{4\pi e}{k\alpha}t$$

$$= \frac{2\pi e^2 t}{k\alpha} \quad\dots\dots\dots\dots\dots\dots(2).$$

Now by equation (1) we see that, if the potentials be kept constant, by connecting the plates to the poles of a battery or in some similar way, the energy of the system is inversely as the distance between the plates, and will therefore be increased if the distance be diminished. The attracting force between the oppositely charged plates tends to diminish the distance between them, but, if the potentials be kept constant, the system draws the necessary energy from the battery.

On the other hand, if the plates be insulated, so that the charges are constant, the system has no source of energy to draw upon, and thus the action which tends to occur under the natural forces, namely, an approach of the plates, must involve a decrease of available energy. This is also seen from equation (2), which shows that, in this·case, the energy diminishes with the distance between the plates.

In the first case, where the potentials are kept constant, the effect on the energy of a small displacement of one plate towards the other is the difference between the energies before and after the displacement. Before the displacement, the energy is $\dfrac{ak(V_1 - V_2)^2}{8\pi t}$, or writing $V$ for $V_1 - V_2$, the energy is $\dfrac{akV^2}{8\pi t}$. After the displacement, the energy is $\dfrac{akV^2}{8\pi(t - \delta t)}$, where $\delta t$ denotes a small change in the thickness $t$ of the dielectric.

Since $\delta t$ is a small quantity, this expression is equivalent to $\dfrac{akV^2}{8\pi t}\left(1 + \dfrac{\delta t}{t}\right)$, and the increase in the energy of the system during the displacement is $\dfrac{akV^2\delta t}{8\pi t^2}$.

Putting $V = 4\pi\sigma t/k$, and $\sigma = e/a$, we see that this expression becomes $\dfrac{2\pi e^2 \delta t}{ak}$.

Now in the second case, when the charges are constant, the energy is $2\pi e^2 t/ka$, and, after the same displacement, it becomes $2\pi e^2(t - \delta t)/ka$. Thus the decrease in energy of the system is $2\pi e^2 \delta t/ak$. It follows that for a small displacement of the kind indicated, the increase of electrical energy, when the potentials are kept constant, is equal to the decrease of energy when the charges

are constant, and this, by the principle of the conservation of
energy, when the system has no actual energy to draw upon, is
equal to the external work which the system can do during the
displacement. A general investigation shows that this result holds
good for any kind of small displacement to which the system is
subjected.

**20.** For many purposes, the best and most sensitive form
The quadrant      of electrometer is the quadrant instrument invented
electrometer.      by Lord Kelvin. A light flat needle of aluminium
or silvered paper is suspended by a fine wire or quartz fibre
within a shallow box divided into four quadrants. The quadrants are supported on insulating pillars, which, in one recent
form of the apparatus, are made of amber. Opposite quadrants
are connected together by fine wires, and, in its position of
equilibrium, the needle lies over the junction line between two
quadrants.

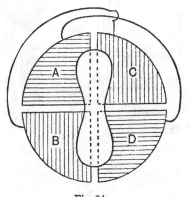

Fig. 24.

The needle is charged to a high potential, and, if there be no
difference of potential between the quadrants, the needle still lies
in its median position. If, however, the opposite pairs of quadrants
be connected with the terminals of a voltaic cell, or other source
of potential-difference, the needle is deflected towards that pair of
quadrants with the lowest potential, that is, with a relative electric

charge opposite to that on the needle. A mirror is attached to
the needle, and the deflection measured by one of the usual
optical methods.

In the form of electrometer shown in Fig. 24 a, the design of
which is due to Dr F. Dolezalek, the suspension is a very fine quartz
fibre, and is made into a conductor by coating it with a trace of a
hygroscopic substance like calcium chloride, which always remains

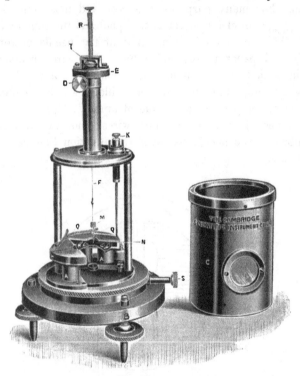

Fig. 24 a.

moist. The needle is of very light silvered paper or thin aluminium
foil, and the sensitiveness is so great that the needle need only be
kept at a potential of 50 to 200 volts. A constant potential can be
secured by connecting the fibre with one pole of a battery or dry
pile, the other pole of which is put to earth. In older forms of the
instrument, in which much higher potentials were necessary, the
needle was charged initially by means of an electric machine or

electrophorus, and was kept at a moderately constant potential by allowing a wire from it to dip in some sulphuric acid placed in a little vessel coated with tin-foil. This vessel served as a Leyden jar, and its large capacity greatly retarded the fall of potential due to leakage from the needle.

The relation between the potentials and the deflection can be found by examining a simpler case consisting of a large plane

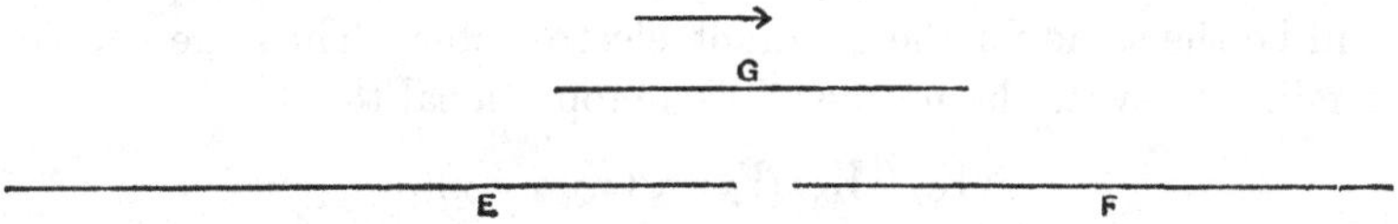

Fig. 25.

surface $G$ moveable in its own plane over the two parallel co-planar surfaces $E$ and $F$. If $l$ be the width of the planes at right angles to the plane of the paper, $Xl$ may be taken to denote the force tending to move $G$ in the direction of the arrow. When $G$ moves through a small distance $x$, the work done is $Xlx$. If the electrical system be isolated, so that the charges are constant, this work is equivalent to the decrease in the electrical energy of the system, while, if the potentials are kept constant, as is more usual in practice, it is equivalent to the increase in the electrical energy of the system (§ 19).

When $G$ is moved through a distance $x$, the area of $G$ opposite to $F$ will be increased by $lx$, and the energy will be increased by

$$\frac{lx}{8\pi t}(V_N - V_{Q_1})^2,$$

where $V_N$ denotes the potential of the needle $G$, and $V_{Q_1}$ that of one quadrant $F$.

At the same time, the area of $G$ opposite to $E$ is decreased, also by $lx$, and a corresponding decrease of energy occurs equal to

$$\frac{lx}{8\pi t}(V_N - V_{Q_2})^2.$$

Thus the total increase in electrical energy at constant potential is

$$\frac{lx}{8\pi t}\{(V_N - V_{Q_1})^2 - (V_N - V_{Q_2})^2\}.$$

As we have seen, this is equal to the work done, therefore

$$Xlx = \frac{lx}{8\pi t}(V^2_{Q_1} - V^2_{Q_2} + 2V_N V_{Q_2} - 2V_N V_{Q_1}).$$

Thus $\qquad X = \frac{1}{4\pi t}(V_{Q_2} - V_{Q_1})\{V_N - \tfrac{1}{2}(V_{Q_1} + V_{Q_2})\}.$

In this expression, the constant $1/4\pi t$ depends on the particular case used for calculation, but the general form of the expression will be the same for the quadrant electrometer. Thus the couple tending to twist the needle will be proportional to

$$(V_{Q_2} - V_{Q_1})\{V_N - \tfrac{1}{2}(V_{Q_1} + V_{Q_2})\}.$$

In practice, the potential of the needle is usually very high compared with that of the quadrants, and the couple will be proportional to

$$(V_{Q_2} - V_{Q_1})V_N,$$

that is, to the potential of the needle and to the potential-difference between the quadrants.

This couple is opposed by the torsional couple in the wire or quartz fibre. The torsional couple is proportional to the angular deflection $\theta$. Thus

$$\theta = kV_N(V_{Q_2} - V_{Q_1}),$$

where $k$ is some constant, to be found by experiment, depending on the form and dimensions of the instrument.

Beattie has given a more complete theory which takes into account the air-space between the quadrants. This explains the fact that the sensitivity is not proportional to the needle-potential, but rises to a maximum and then falls.

Quadrant electrometers are often used to measure large differences of potential by connecting one pair of quadrants with the needle. We then have $V_N = V_{Q_2}$ and the general expression becomes

$$(V_N - V_{Q_1})\{V_N - \tfrac{1}{2}V_{Q_1} - \tfrac{1}{2}V_N\} = \tfrac{1}{2}(V_N - V_{Q_1})^2.$$

Thus the deflection will be proportional to the square of the difference of potential between the needle and the other pair of quadrants.

This principle is employed in the electrostatic voltmeter illustrated in Fig. 26. Here two of the quadrants are removed, the axis of the needle is horizontal, and the lower end of the needle carries an adjustable weight. The needle is connected with the quadrants, and thus both are raised to the same potential.

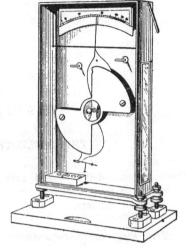

It is obvious that the result is the same whether the system be positively or negatively electrified. Hence the electrostatic voltmeter, unlike some other forms, can be used to measure an alternating difference of potential.

In the multicellular voltmeter, the sensitiveness is in-

Fig. 26.

creased by attaching a number of needles to a vertical axis, while quadrants lie between the needles.

Ayrton and Mather introduced another form of electrostatic voltmeter, in which the "needle" takes the form of two cylindrical plates, hanging vertically from a phosphor-bronze strip, and attracted when charged into two cylindrical inductors, which play the part of the quadrants in the other forms of electrometer. The oscillations of this voltmeter are of short period and are quickly damped.

Methods of using these electrostatic voltmeters to measure current and voltage are described in the chapter on the electric current, p. 134.

the chapter on the electric current, p. 134.

### REFERENCE.

*Elements of the Mathematical Theory of Electricity and Magnetism;*
by Sir J. J. Thomson; Chapters I, III, V.

# CHAPTER III

## THE DIELECTRIC MEDIUM

The importance of the dielectric medium. Lines and tubes of force. The energy in the dielectric medium. Analogy with a strained medium. Dielectric currents.

**21.** THE conspicuous success of Newton's formulation of the laws of gravitation suggested similar relations for electric forces, and, when the law of inverse squares was verified for electric forces also, it was inevitable that the analogy should be pushed as far as, or farther than, the case warranted. The gravitational attraction between two masses is independent of the nature of the intervening medium; and, forgetful or in ignorance of the experiment of Cavendish to which we have referred in § 11, natural philosophers till the time of Faraday assumed that electric forces also were not affected by the insulating medium across which they acted.

With an obstinate disbelief in the idea of action at a distance, Faraday set himself to examine the influence of the dielectric field. Guided by the hypothesis that the forces were somehow transmitted by the medium, he rediscovered the existence of specific inductive capacity, and, as we have seen, measured its value for several substances. He also showed by many experiments that electrostatic induction might occur in curved lines, whereas, on the theory of direct action at a distance, it was supposed to act in straight lines only. As a final result of his work, he framed a new theory of electric phenomena. He regarded them all as depending "on induction being an action of the contiguous particles of the dielectric, which, being thrown into a state of polarity and tension, are in mutual relation by their forces in all directions."

Faraday's experimental researches in these subjects not only confirmed his original hypothesis, but still serve as an excellent example of the advantage, perhaps the necessity, of hypothesis as an aid to physical investigation even in the early stages of a course of experiment.

The centre of interest in electric science was thus shifted from the conductors to the dielectric or insulating medium. In Faraday's eyes, the essence of the phenomena was to be found in a state of strain set up by the action of the electric machine in the dielectric, the so-called electric charges on neighbouring conductors being merely the free surfaces of the strained medium.

**22.** In order to study the phenomena of the dielectric field,
Lines and tubes of force. Faraday made use of the ideas of lines and tubes of force, conceptions which seem to have been suggested by the pattern assumed by iron filings when scattered on a horizontal card in a magnetic field (Fig. 27). Each filing, under the

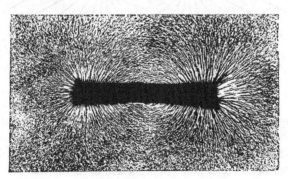

Fig. 27.

inductive action of a magnet placed below the card, becomes a little temporary magnet; the filings set, each in the direction of the local magnetic force, and cling together to form chain-curves, the tangent to a curve at any point being in the direction of the horizontal magnetic force at that point.

The corresponding lines of electric force, due to electrostatic charges, cannot be demonstrated experimentally in such a simple manner, but their forms can be calculated from the law of force, and thus maps of the lines constructed for any simple case. A line of force is to be defined as a curve drawn so that, at each point, the tangent to the curve is in the direction of the resultant electric intensity at the point.

Two charges, equal in amount but opposite in sign, produce a simple field of force. Placed between them, an isolated unit point-charge of positive electricity will be repelled by one and attracted by the other. One line of force, then, must run straight from one of our charges to the other. At a point in the median plane, when

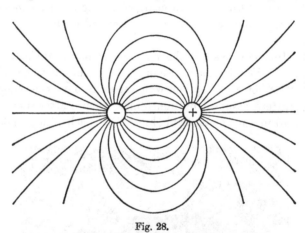

Fig. 28.

the charges lie in the same horizontal line, the vertical components due to the two charges will balance each other, and the horizontal components will again be in the same direction. Thus all lines of force cross the median plane in a horizontal direction. If we make the convention that a single line of force shall spring from one positive unit of electricity and end on one negative unit, the number of lines crossing any small area in the dielectric field normal to the lines will, as we shall see, be proportional to the resultant force at the area. We then arrive at the general picture of the lines shown in Figure 28, in which the resemblance to the corresponding case for magnetic lines is shown.

Figure 29 shows the distribution of the lines for two opposite charges, $A$ being four times as great as $B$. Here only part of the lines springing from $A$ end on $B$, the rest going to the walls of the room, or other conducting surfaces on which a negative charge can be induced.

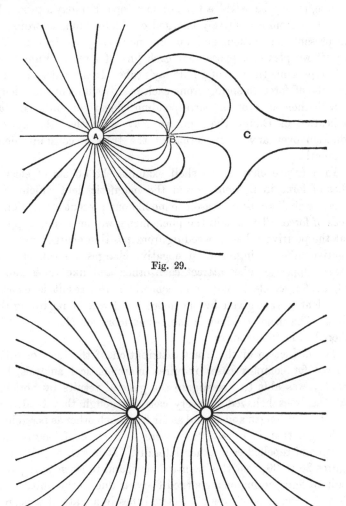

Fig. 29.

Fig. 30.

In Figure 30 the lines of force arise from two equal positive
charges of the same name, and all end on surrounding objects.
At the point marked $C$ in Figure 29, and at a point midway
between the charges in Figure 30, there is no resultant force.

If a tubular region of space be imagined as bounded by lines
of force, it may be called a tube of force or a Faraday's tube.   It
must start from a positively charged conductor, and, according to
our present convention, end on one negatively electrified.   We
may, if we please, suppose that each line of force in Figures 28
to 30 represents the axis of a Faraday's tube.   It is usual to imagine
one tube of force to spring from each positive unit of electricity.
It will then end on a corresponding negative unit.   By an
application of Gauss' theorem, we may show that the normal
induction over any cross-section of the tube is constant along
its length.

In a future chapter we shall examine the effects of electric
tubes of force in motion.   From the magnetic field which then
arises, we shall see some reason to modify our present conception of
tubes of force.   To explain the phenomena, we must then suppose
that the positive tubes proceeding from positive charges, and the
negative tubes springing from negative charges, all run off into
space.   Opposite tubes attract each other, and like tubes repel;
and, as far as electrostatics is concerned, the result is exactly
equivalent to that given by the more usual way of regarding the
tubes.   The usual method is somewhat more simple, and will be
adhered to in this chapter.

The deduction of the exact distribution of the lines or tubes
of force for complex cases requires mathematical analysis, but
Faraday was able to investigate the general results by noticing
that the lines behaved in every case as though they tended to
shorten their length and spread as far from each other as possible—
as though, that is to say, they were in a state of longitudinal
tension and repelled each other laterally.   It is well to reconsider
Figures 28 to 30 from this point of view, and to learn to regard a
resultant force, acting on a charged body, as due to the unbalanced
effect of a one-sided excess of electric tubes in a state of tension.

We proved (§ 18) that on each unit area of a charged conductor
there existed a mechanical force equal to $f\sigma/2$, $f$ being the

resultant electric intensity, and $\sigma$ the surface density of electrification. Now each line or tube of force is drawn from one unit charge, so that $\sigma$ denotes also the number of tubes of force proceeding from unit area. Hence the mechanical force on the surface is the same as though each tube exerted a pull equal to $f/2$. Thus the mechanical forces in the field are the same as though the tubes of force were in a state of tension, the tension in a tube at each point being measured by one-half the electric intensity at that point.

The electric intensity $f$ just outside a charged surface is $4\pi\sigma/k$, and, since $\sigma$ also denotes the number $N$ of tubes of force per unit area, the electric intensity at any point of the field is $4\pi N/k$.

Thus
$$N = \frac{fk}{4\pi}.$$

Hence the tension due to the tubes of force across unit area is

$$\tfrac{1}{2}Nf = \frac{f^2k}{8\pi}.$$

If the tensions were the only stresses in the field, the tubes of force between two oppositely electrified curved conductors would all tend to gather together and run in straight lines from the positive to the negative charges by the shortest path. In order to represent completely the state of the dielectric medium, in cases where the tubes of force are not straight, we must also suppose that the tubes repel each other, so that a lateral pressure exists at right angles to the lengths of the tubes. It may be shown by mathematical analysis that the requisite pressure also has the numerical value $f^2k/8\pi$, which is equal to the tension along the length of the tubes per unit area of cross-section.

**23.** As an example of a simple system, let us consider two parallel planes charged with equal and opposite quantities of electricity. The total energy is, according to § 19,

*The energy in the dielectric medium.*

$$W = \tfrac{1}{2}e(V_1 - V_2),$$

where $e$ is the charge on one plane, and $V_1 - V_2$ the difference of potential between the planes.

Since the electric intensity $f$ between the planes is uniform,

$$V_1 - V_2 = ft,$$

where $t$ is the distance between them. Now, if $\sigma$ be the surface density, $f$ is $4\pi\sigma/k$, and $\sigma$ is therefore $fk/4\pi$. Thus the energy

$$W = \tfrac{1}{2}\sigma A ft$$

$$= \frac{f^2 k}{8\pi} . At.$$

But $At$, the area of one plane multiplied by the distance between the planes, represents the volume of the dielectric stratum. Thus the energy per unit volume is $f^2 k/8\pi$; and this expression has a form which is independent of the particular limitations of the system chosen. At any point in a dielectric field, then, the energy residing in a surrounding volume, small enough for the intensity throughout it to be uniform, is $f^2 k/8\pi$ per unit volume.

**24.** An elastic mechanical system, such as a spiral spring stretched by a load, possesses potential energy when

Analogy with a strained medium.

strained, and one of the essential features of the electric theory, as formulated by Faraday and developed mathematically by Clerk Maxwell, consists in tracing the analogy between a dielectric medium subject to electric forces and a medium strained mechanically.

An elastic system, where the force $F$ is proportional to the displacement $x$, possesses potential energy equal to the work done in stretching, which is $\tfrac{1}{2}Fx$. The expression for the electric energy per unit volume of insulator, $f^2 k/8\pi$, may be written as

$$\tfrac{1}{2}f . \frac{fk}{4\pi},$$

when its analogy with the mechanical energy

$$\tfrac{1}{2}F . x$$

is manifest. Corresponding to $f$, the electric force, we have $F$ the mechanical force, and corresponding to $x$, the mechanical displacement, we have the quantity $fk/4\pi$, which, on this analogy, was called by Maxwell the electric displacement, though it is perhaps better described by Faraday's older name of dielectric polarization.

The view which locates the energy of an electric system in the dielectric medium must be regarded as one of the most important and fundamental conceptions in modern physical science. The germ of the idea is to be found in Franklin's work, by which he showed that the charge of a Leyden jar or other condenser resided in the glass or other insulator which separated the two coatings. Cavendish, too, clearly recognised the effect of the dielectric on the capacity of his condensers.

But it is to the instinctive genius of Faraday that we owe the first definite formulation of the new theory. As we have already stated, Faraday regarded induction as " an action of the contiguous particles of the dielectric, which, being thrown into a state of polarity and tension, are in mutual relation by their forces in all directions." It is not too much to say that this hypothesis was the guiding star in all Faraday's researches in electrostatics. The wonderful relations between apparently unconnected phenomena discovered by Faraday, Maxwell, and their followers, are the best evidence in favour of the trustworthiness of their faith.

Faraday had no facility in mathematical analysis, and his ideas, though fruitful indeed in his own case, only dominated the work of others when translated into definite mathematical form by his great successor Clerk Maxwell. If, as indicated above, the energy of the electric system resides in the dielectric medium, there should be a finite time required for electric induction to pass from one body to another—a time, during which, short though it may be, the energy is passing through the dielectric medium, and is unconnected in any way with conducting bodies. Such ideas led Maxwell to the theory of electromagnetic waves, and enabled him to calculate their velocity by methods we shall study in a later chapter. The concordance of the theoretical value of this velocity with the observed velocity of light, led directly to the theory that light is a series of electromagnetic waves in the dielectric luminiferous medium.

These ideas we shall examine in future chapters; here we are concerned chiefly with electrostatic phenomena. But, even in these phenomena, Faraday's views lead to modifications in our fundamental conceptions. Instead of fixing our attention on the charges, and forcing us to frame theories of incompressible fluids to explain the ideas in our minds, Faraday asks us to ignore the charges, and seek an explanation of electrostatic manifestations in the conception of energy residing in the dielectric medium, by virtue of stresses and strains therein. The so-called charges on conductors in the electric field are but the free surfaces of the strained intervening substance. The charges serve to demonstrate the state of strain, as a spring balance, inserted into the middle of a stretched india-rubber cord, may be used to show the tension.

Electric forces can act across a vacuum, hence material media are not necessary. We are led to refer the stresses and strains here described to some all-pervading medium, which may be modified, but not excluded, by the presence of ordinary matter. We shall see later that the medium it is necessary to conceive in order to explain electric and magnetic phenomena is identical in properties with the medium required to explain light.

Faraday's tubes of force, and Maxwell's electromagnetic equations, fixed the centre of interest in the dielectric, till the work of the last few years has made it necessary to ask what happens at the ends of the tubes of force, and again to examine into the nature of the electric charges. To such points we shall return later.

**25.** Fig. 31 represents the system of tubes of force which
Dielectric
currents.
connect the opposite plates of a parallel plate condenser. Most of these tubes run straight from one

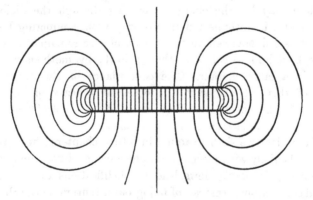

Fig. 31.

plate to the other, but, near the edges, tubes leak out, and some few join the back of the plates.

Now let us suppose that the two plates are connected together suddenly by means of a metallic wire. In a minute fraction of a second the condenser is discharged, and the whole system of tubes of force, representing the state of strain in the dielectric medium, vanishes.

In order to trace the process of their disappearance, three things must be borne in mind: (1) the tubes are in a state of tension, and thus tend to contract; (2) they repel each other laterally; (3) they end on conductors, the free ends of the tubes representing charges of positive and negative electricity.

While the condenser remains charged, the tubes of force may be regarded as in equilibrium under their own tensions and pressures. When, however, a metallic connexion between the plates is established, the charges near the ends of the wire are free to move. That is to say, the ends of the tubes which are anchored to the charges (or, rather, which themselves constitute the so-called charges) drag their anchors. A tube of force with its ends at different points on the same wire will contract under its own tension, and its opposite ends will approach each other. The tube will shut up, and be obliterated.

The disappearance of one tube leaves the remaining tubes in a state which may be regarded as one of unstable equilibrium; the lateral pressure of that tube is removed, and neighbouring tubes will be pushed by the unbalanced pressure of other tubes successively into the wire. In this way the dielectric field is relieved of strain, and the tubes of force are destroyed.

During this process a current of electricity is said to flow along the conducting wire. On the view we are now considering, that current is represented by the process of the tubes of force dragging the charges to which their ends are anchored along the wire under the tension of the tubes. The heat developed by the current is the result of the friction produced by the resistance of the conductor to the drag of these anchors.

In a later chapter we shall see that an electric current possesses something analogous to inertia, and that the process of discharge sometimes consists of a series of oscillations. The current alternates, and passes first in one direction and then in the other; it diminishes gradually in intensity till it dies away and the charges disappear.

The electric current across any surface is measured by the amount of charge passing across it per second, and, considering unit area of the planes, the current entering it is measured by the rate of change of $\sigma$ the charge per unit area. Now, as we saw on pp. 38, 39, $\sigma$ is equal to $fk/4\pi$, which itself (p. 56) represents the strain in the dielectric medium, or the dielectric polarization.

Thus an electric current is represented by a change in the dielectric polarization, and a circuit such as that we have been considering is completed by the dielectric medium. On the older view, the process of charging the condenser is imagined as the flow of the opposite charges along the two wires on to the plates, somewhat as two separate streams of water might be supposed to flow along two pipes into two reservoirs. On the theory of Faraday and Maxwell, we must imagine this process to be similar to a transient current flowing round a closed circuit till, owing to the strains in the dielectric medium, the opposing electromotive force set up in one part of the circuit becomes sufficient to produce equilibrium. In this part of the circuit no slip occurs, and all the energy put in remains, except for radiation, as the potential energy of dielectric strain. In the wires, slip occurs with consequent friction, and the energy which enters sets up no opposing strain: it is all dissipated in the friction of electric resistance, and appears as heat in the conductor.

## REFERENCES.

*Experimental Researches in Electricity*; by Michael Faraday.
*Elementary Treatise on Electricity*; by J. Clerk Maxwell; Chapters IV, V, IX.
*Electricity and Magnetism*; by J. Clerk Maxwell; Chapters I, II, V.
*Recent Researches in Electricity and Magnetism*; by Sir J. J. Thomson; Chapter I.
*Elements of the Mathematical Theory of Electricity and Magnetism*; by Sir J. J. Thomson; Chapters II, IV.
*Electricity and Matter*; by Sir J. J. Thomson.

# CHAPTER IV

## MAGNETISM

Early history. Principal magnetic phenomena. Variation of magnetic force with distance. Interaction of two short magnets. Magnetic potential. Terrestrial magnetism. Magnetic induction. Experiments on magnetic induction. Theories of magnetism.

**26.** THE lodestone (or leading stone), a magnetic oxide of iron, was known to the ancients, and, since it was first found near the town of Magnesia in Lydia, received the name μάγνης. Lucretius appears to have known that the lodestone attracts iron, and that iron itself, when in contact with a lodestone, acquires magnetic properties.

Early history.

The first practical application of magnetic properties was the invention of the mariner's compass, and, as so often happens, the interest aroused by a practical application led to a great development of the theoretical side of the subject.

It has been stated that the magnetized steel needle or mariner's compass was early known to the Chinese, but, however this may be, descriptions of it seem first to have appeared in European literature about the 12th century, coming probably from Saracenic sources, though the references indicate that it had been known for some time.

A compass-needle points not directly north and south, but in a magnetic meridian making an angle, called the declination or variation, with the geographical meridian.

Stephen Burrowes discovered that this angle changed in the course of a voyage; and, in 1683, Halley showed that the differences in the declination observed at sea were not due to the effect of neighbouring land, but must depend on the magnetic properties of the earth as a whole.   In 1698 he was appointed captain of an exploring ship, and, with this command, undertook a voyage which lasted two years, for the purpose of elucidating the phenomena of terrestrial magnetism.   Halley, who afterwards succeeded Flamsteed as astronomer royal, also pointed out the importance of the slow secular change in the declination, which between 1580 and 1692 had changed from 11° 15' east to 6° west.   The existence of smaller variations, both annual and daily, was noted by Graham about 1719.

About 1544 Georg Hartmann, vicar of St Sebaldus, Nuremberg, observed that a needle, pivoted to move freely in a vertical plane, dips down towards the north when magnetized by a lodestone.   This dip or inclination was rediscovered in 1576, and investigated carefully by Robert Norman, a mariner and hydrographer.   Norman found that at London the angle of dip was 71° 50'.   This angle too is subject to periodic change.

From 1540 to 1603 lived William Gilbert, of Colchester, who, besides the electrical researches to which we have referred in the first chapter, in his great work *De Magnete*, collected all that was then known on the subject of magnetism, and added many new and valuable observations.   To Gilbert we owe the conception of the earth as a huge magnet, and the first exact studies which virtually founded magnetism as a science.

**27.**   A steel needle, magnetized by stroking with a lodestone, and pivoted so as to be free to turn horizontally, sets in the magnetic meridian—its ends point, one towards the north and one towards the south.

Principal magnetic phenomena.

With two pivoted compass-needles, it is easy to show that two north-seeking or two south-seeking poles repel each other, while two unlike poles attract each other.   Here there is an obvious analogy with electric phenomena, but, unlike electric phenomena, magnetic effects are observed to any great extent with a few substances such as iron only, and act with equal intensity across the intervening medium whatever be its nature, provided it is not of iron, nickel or other magnetic substance.   Screens of paraffin, ebonite, copper and all other non-magnetic bodies, produce no change in the force with which one magnet acts on another, though, as we have seen (p. 20), such screens diminish or destroy electric action. Again, an electric conductor can be given an isolated charge of positive or negative electricity, but a magnet must acquire an equal south-seeking pole when north-seeking magnetism is impressed upon it. In this respect, the substance of the magnet is more nearly analogous to the dielectric medium, at the boundaries of which equal and opposite electric charges are always to be found.

Just as electric charges are induced on conductors by the presence of charges in their neighbourhood, so magnetic poles appear on pieces of iron placed near a permanent magnet. A permanent north-seeking pole will induce a south-seeking pole on the nearer portions of a block of iron and a north-seeking pole on the further portions. This explains the attractions observed between a permanent magnet and pieces of iron originally un-magnetized. Iron filings, for instance, will cling to the poles of a magnet and even form long chains: poles are induced on each filing and the opposite poles of neighbouring filings cling together.

A difference in behaviour between soft iron and steel becomes apparent at this stage of our experiments, and is of importance. The readily-induced magnetism on soft iron is temporary. If the inducing magnetic force be removed, the induced magnetism is destroyed by a shake or shock, or by a small magnetic force in the reverse direction. With hard steel, on the other hand, while the induced magnetization is less intense for a given mag-netizing force, it is much more permanent, and a considerable fraction of its intensity will survive even rough usage. Heating, however, will destroy all magnetization in any case. At a critical temperature of about 700° to 900° Centigrade, both iron and steel cease to be magnetic substances. On cooling they regain their magnetic properties, but any actual magnetism they possessed before the heating will have disappeared.

The clinging filings to which we have referred indicate clearly the points from which the forces originate, that is, the poles of the magnet. The poles are not situated quite at the ends

of the bar, but at places some little distance from the ends. The
line joining these poles is known as the magnetic axis, and, for
a thin needle, coincides with the geometrical axis. A magnetized
mass of steel of whatever shape, however, still possesses a magnetic
axis, which, if the mass were free to turn, would lie always in the
direction of the resultant magnetic force.

Like the lines of electric force, magnetic lines of force may be
defined as curves drawn through the magnetic field, so that, at
each point of their length, the tangent to the curve gives the
direction of the resultant local magnetic force. If filings are
sprinkled over a card, below which a magnet is placed, pictures
of the lines of force in the plane of the card are readily obtained.
A figure showing these lines has already been given to illustrate
the idea of electric lines of force, and will be found on p. 51
(Fig. 27).

**28.** As in the case of gravitation and electric force, the force
which a single magnetic pole will exert on another
pole is found by experiment to vary inversely as the
square of the distance. So general is this form of
law for all kinds of force acting from a point-source,
that, from this group of experiences alone, we might almost
conclude that the inverse square law is a general property of
our conception of space, rather than a relation depending on
the nature of the particular forces.

*Variation of magnetic force with distance.*

The knowledge of the law of force enables us to define our
magnetic pole of unit strength as the pole which, when surrounded
by air and placed at unit distance from an equal similar pole,
repels it with unit force. The magnetic force at a point is then
defined as the mechanical force on a pole placed at the point
divided by the strength of the pole.

A modification of the torsion balance enabled Coulomb approximately to establish the law of inverse squares in the case
of magnetic poles, as well as in that of electric charges. The
disturbing effects of the opposite poles, which must necessarily
exist, and can only be removed to moderate distances, make the
experiment somewhat inaccurate, and has led to the development
of better methods.

All such methods depend on the measurement of the magnetic
force at a point by means of the
deflection of a magnetic needle from
its normal position of equilibrium.
Let *NS* denote the magnetic meridian of the earth and let a magnetic
force *F* be applied along the magnetic
east and west line, at right angles
to the horizontal component of the
earth's magnetic force, which we will
denote by *H*.

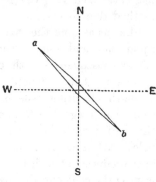

Fig. 32.

The magnetic needle *ab* is thus
deflected into the position shown,
at an angle $\theta$ with the meridian.
If *m* be the strength of each pole, the horizontal force on it due
to the earth is *Hm*, and if *l* be the half distance between the
poles of the needle, the couple due to the earth is $2Hml \sin \theta$.
The couple due to the applied magnetic force *F* is $2Fml \cos \theta$.
The quantity $2ml$ constantly recurs in magnetic problems, and
is known as the magnetic moment of the magnet. We shall
denote it by the symbol *M*. It may be defined as the strength
of one pole multiplied by the distance between the poles, and
it is clearly measured by the couple exerted on the magnet
when placed at right angles to the magnetic force in a field of
unit intensity.

When the magnetic needle we are considering is in equilibrium, we get

$$FM \cos \theta = HM \sin \theta,$$

or
$$F = H \tan \theta.$$

It will now be clear that we can compare two magnetic forces
by comparing the deflections of the same compass-needle produced
successively by the forces. For in each case

$$F = H \tan \theta.$$

If *H* be the same, the forces will be proportional to the tangents
of the angles of deflection.

The most accurate verification of the law of force due to a
magnetic pole is given by a
method due to Gauss.

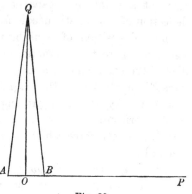

Let us assume the inverse
square law, and compare some
of its consequences with experiment. Let $m$ denote the
strength of either pole of a
short bar magnet $AB$, and $l$
the half length between its
poles. At a point $P$ on its
axis produced, at a distance $r$
from $O$ the centre of the magnet, the total force

Fig. 33.

$$F_1 = \frac{m}{(r-l)^2} - \frac{m}{(r+l)^2} = \frac{m\{(r+l)^2 - (r-l)^2\}}{(r-l)^2(r+l)^2}$$

$$= \frac{4mlr}{r^4 - 2r^2l^2 + l^4}$$

Neglecting terms involving the squares and higher powers of the
comparatively small quantity $l$, we have

$$F_1 = \frac{2Mr}{r^4} = \frac{2M}{r^3}$$

Now let us calculate the force at a point $Q$ on the normal to
the axis through the centre of the magnet. Let $OQ$ be made
equal to $OP$ in the former experiment, and be denoted by $r$. The
forces due to the two poles act along $AQ$ and $BQ$ respectively, and
the resultant evidently acts from $Q$ parallel to the axis of the
magnet $AB$. The value of this force is the sum of the resolved
components; thus

$$F_2 = 2\frac{m}{(AQ)^2} \cdot \frac{OA}{AQ} = 2\frac{m}{r^2 + l^2} \cdot \frac{l}{(r^2 + l^2)^{\frac{1}{2}}}$$

$$= \frac{2ml}{(r^2 + l^2)^{\frac{3}{2}}} = \frac{M}{r^3\left(1 + \frac{l^2}{r^2}\right)^{\frac{3}{2}}}.$$

Again neglecting terms involving $l^2/r^2$, we have

$$F_2 = \frac{M}{r^3}.$$

It follows, then, as a consequence of the law of inverse squares
for the force due to a magnetic pole, that the force produced by a
short magnet in the first, or so-called end-on position, is double
the force in the second or broad-side-on position.  Moreover, in
each position, the force due to the whole magnet varies inversely
as the cube of the distance from its centre.  Both these results,
it is evident, depend on the assumption that the force due to a
single pole is that of the inverse square.  If the force of a single
pole varied as the inverse $n$th power of the distance, it is easily
seen that the force due to the whole magnet, when it is short,
would diminish in the ratio of the $(n+1)$th power, and the force
in the end-on position would be $n$ times that in the broad-side-on
position.  We should have

$$F_1 = \frac{nM}{r^{n+1}} \text{ and } F_2 = \frac{M}{r^{n+1}}.$$

In order to compare the magnetic forces produced by a magnet
at equal distances in the two standard
positions, a light mirror, with three or
four pieces of magnetized watch-spring
fixed behind it, may be suspended in a
small glass case in the earth's field by a
fine fibre of unspun silk.  Its position of
equilibrium is noted by reflecting a spot
of light on to a scale.  A short bar magnet
is then placed successively east and west
of the magnetometer mirror in the end-on
position, and the mean deflection measured.
The magnet is then placed with its centre

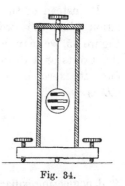

Fig. 34.

successively north and south of the mirror in the broad-side-on
position.  The distance from the centre of the magnet to the
centre of the mirror is in each case made as nearly as possible
the same.  It will be found that the tangent of the angular
deflection in the first position is very nearly indeed twice the
tangent of the angular deflection in the second position.  Moreover, by changing the distance, it will be found that the tangent
of the deflection varies inversely as the cube of the distance.
We thus obtain an independent verification of the law that the

magnetic force of a pole varies inversely as the square of the distance.

**29.** For the sake of example, let us consider two short magnets in the end-on position relatively to each other. Let $r$ be the distance between their centres. The magnet $AB$ produces a magnetic force $2M/r^3$ at the part of the field where lies the magnet $CD$, and, if $M'$ be the

*Interaction of two short magnets.*

Fig. 35.

moment of the magnet $CD$, the couple acting on it when deflected through an angle $\theta$ is clearly

$$C = \frac{2MM'}{r^3} \sin \theta.$$

If, instead of the couple, we wish to find the force of translation when the two magnets lie end-on to each other as in Figure 35, we may proceed as follows. The magnetic force which $AB$ exerts at a distance $r$ from its centre is $2M/r^3$. The total force it exerts on the two poles ($m'$ and $-m'$) of the magnet $CD$ is

$$F = \frac{2Mm'}{(r - l')^3} - \frac{2Mm'}{(r + l')^3}$$

$$= 2Mm' \frac{(r + l)^3 - (r - l)^3}{(r - l)^3 (r + l)^3}$$

And, neglecting squares and higher powers of the small quantity $l$,

$$F = 2Mm' \frac{6r^2l'}{r^6} = 6M \frac{2m'l'}{r^4}$$

$$= \frac{6MM'}{r^4}.$$

Thus the forces vary inversely as the fourth power of the distance, while the couples vary inversely as the third power. These results hold whatever be the relative positions of the two magnets, though of course the absolute values of the couples and forces depend on the inclination of the magnets to each other. Similar investigations may be made for other positions.

**30.** The difference in magnetic potential between two points
**Magnetic** is defined as the quantity of work required to carry
**potential.** an isolated north-seeking magnetic pole of unit
strength from one point to the other. The absolute potential of a point is then the amount of work required to bring such a pole against the magnetic forces from a place of zero potential (or from infinite distance) to the point considered, on the assumption that the strength of the pole is small enough not to disturb the magnetic field.

The investigation given in § 8 for the electric potential of a point in air, at a distance $r$ from an electric charge $e$, applies, with the necessary change of symbols, to the case of the magnetic potential of a point in air or any other non-magnetic medium. The magnetic potential is $m/r$, where $m$ is the strength of an isolated magnetic pole at a distance $r$ from the point.

To find the potential due to a short complete magnet, we must consider the combined effect of its two poles at a point $P$ outside it. Join $P$ with $O$ the centre of the magnet, and let the angle between $OP$ and $AB$, the magnetic axis of the magnet, be $\theta$. From $A$ and $B$ draw straight lines $AM$ and $BN$ at right angles to $OP$ or $OP$ produced.

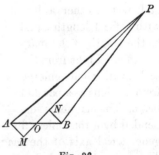

Fig. 36.

The potential at $P$ due to the pole $B$ is $m/BP$, and that due to the pole $A$ is $-m/AP$. Thus, if $OP$ be denoted by $r$, and $OA$ or $OB$ by $l$, the potential $V$ at $P$ due to the magnet is given approximately by

$$V = \frac{m}{PN} - \frac{m}{PM} = \frac{m}{r - l\cos\theta} - \frac{m}{r + l\cos\theta},$$

or, neglecting terms involving the squares of the small quantity $l$,

$$V = \frac{2ml\cos\theta}{r^2} = \frac{M\cos\theta}{r^2}.$$

**31.** The investigation of the distribution of the earth's magnetic force can now be described. The total magnetic
force, at any place on the earth's surface, acts at a
certain angle with the horizontal known as the dip, and at a
certain angle with the
geographical meridian
called the declination. The
total magnetic force can
be determined in magni-
tude and direction by
measuring these two an-
gles and the absolute value
of the horizontal compo-
nent.

Terrestrial magnetism.

Figure 37 shows a de-
clinometer. The magnet
consists of a hollow steel
tube, at one end of which
is placed a piece of plane
glass with a scale on it,
and at the other a lens
with a focal length equal
to the length of the mag-
net-tube. Light from the
scale will then emerge
from the lens as a parallel
beam. The magnet is sus-

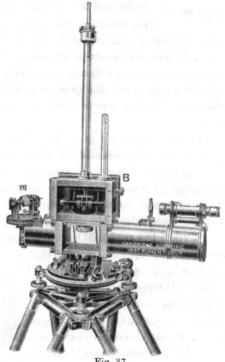

Fig. 37.

pended by a torsionless fibre of unspun silk. The position of the
geometrical axis of the magnet is determined on the glass scale.
The magnet is then turned upside down and resuspended. The
mean of the two positions of the geometrical axis, as observed in
the telescope, gives that of the magnetic axis, and its direction
relative to that of some external object is found by means of the
scale at the foot of the instrument.

A dip-circle is shown in Figure 38. The plane of the circle
is set in the magnetic meridian, and the dip observed. To eliminate
the error arising from the axle of the needle not coinciding with
the centre of the circle, the positions of both ends of the needle

are read through microscopes; to avoid the error due to the magnetic axis not coinciding with the line joining the ends of the needle, the needle is reversed, so that the face originally pointing east now turns to the west; to avoid the error due to the centre of gravity not falling in the line of the axle, the needle is remagnetized, its poles reversed, and a fresh set of observations made. The mean of all these results is taken as the true dip.

Fig. 38.

Both the declination and the dip vary from place to place over the surface of the earth, and the results of such measurements as we have described may be exhibited by maps, on which places of equal declination and places of equal dip are joined by isogonic and isoclinic lines respectively. Another series of lines may be drawn through places of equal horizontal magnetic force. These lines, together with isogonic and isoclinic lines for a date about 1895, are shown in Figure 39.

A different system of directional lines is due to Duperrey, and consists of a series of lines traced by starting from any point and following the direction of the compass-needle.

The slow secular changes in the magnetic angles at London are shown in the table on p. 73.

Besides these slow secular changes, diurnal variations of several minutes of angle are observed. Moreover, rapid and irregular changes, known as magnetic storms, frequently occur, often simultaneously with the formation of a sun-spot, and the occurrence of brilliant auroral displays. Such variations are recorded at magnetic observatories by means of suspended needles, which, through the use of mirrors, are made to record photographically the magnetic changes.

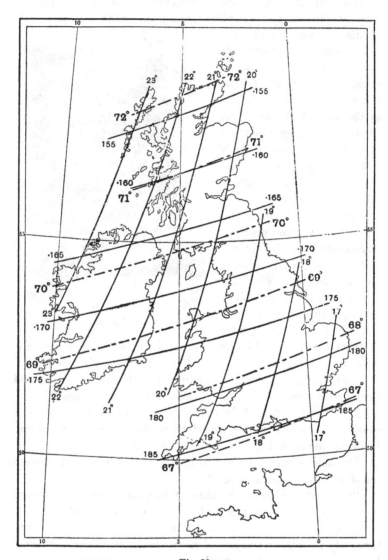

Fig. 39.

| Date | Declination | Dip |
|------|-------------|-----|
| 1576 | | 71° 50′ |
| 1580 | 11° 15′ E. | |
| 1600 | | 72° 0′ |
| 1622 | 6° 0′ E. | |
| 1657 | 0° 0′ | |
| 1672 | 2° 30′ W. | |
| 1676 | | 73° 30′ |
| 1723 | 14° 17′ W. | 74° 42′ |
| 1773 | 21° 9′ W. | 72° 19′ |
| 1787 | 23° 19′ W. | 72° 8′ |
| 1802 | 24° 6′ W. | 70° 36′ |
| 1820 | 24° 34½′ W. | 70° 3′ |
| 1860 | 21° 39′ 51″ W. | 68° 19′ |
| 1893 | 17° 27′ 0″ W. | 67° 30′ |
| 1900 | 16° 52′ 40″ W. | |

By examining the relation which exists between the horizontal and vertical components of the earth's magnetic force, it has been shown that the mean values of the magnetic elements are determined by causes below the surface of the earth, but that the diurnal variation and magnetic storms are due principally to changes going on above the surface—perhaps to electric currents in the rarefied upper regions of the atmosphere.

Having determined the angles of declination and dip, we still require to measure the magnitude of the horizontal component $H$ of the earth's magnetic force, in order to be able to specify completely the magnetic state of a given place.

We showed in § 28 that, at a point on its axis at a distance $r$ from the centre of the magnet, a short bar magnet produced a magnetic force equal to $2M/r^3$. If, then, such a magnet be placed with its axis east and west, it will deflect a small compass-needle placed on its axis through an angle $\theta$, given by the relation

$$\frac{2M}{r^3} = H \tan \theta \ldots\ldots\ldots(1).$$

The angle $\theta$ may be determined accurately by fixing the compass-needle to a mirror, and observing the deflection of a spot of light.

We thus obtain one relation between $M$, the magnetic moment of the deflecting magnet, and $H$, the horizontal component of the earth's force. To get a second relation, the bar magnet may be suspended by a torsionless silk fibre in a glass case. The magnet is then set oscillating through a very small angle about a vertical axis, and its time $T$, of complete oscillation, measured with a chronometer. The restoring couple is $HM \sin \theta$, or, since $\theta$ is small, $HM\theta$. Hence, from the usual formula for harmonic vibrations, we have

$$T = 2\pi \sqrt{\frac{K}{HM}} \quad \dots\dots\dots\dots\dots\dots\dots(2).$$

The moment of inertia $K$ may be calculated from the mass and dimensions of the magnet and its suspension, and we then obtain the value of $H$ from the following expression, which may be deduced from equations (1) and (2) given above

$$H^2 = \frac{8\pi^2 K}{T^2 r^3 \tan \theta},$$

while $M$, the magnetic moment of the bar magnet, is given by the relation

$$M^2 = \frac{2\pi^2 K}{T^2} r^3 \tan \theta.$$

Having thus found $H$, the horizontal component of the earth's magnetic force, by observing $i$ the angle of dip, we know the total magnetic intensity $I$ and the vertical component $V$, for

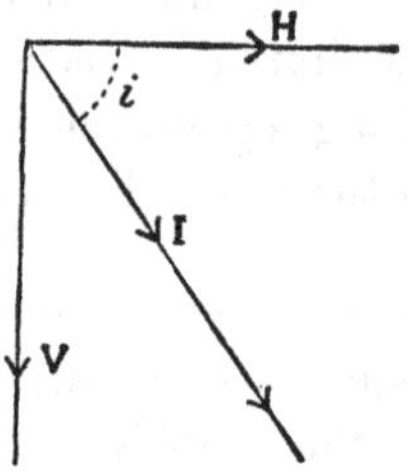

Fig. 40.

$$I = \frac{H}{\cos i},$$

and

$$V = I \sin i = H \tan i$$

**32.** It is well known that magnetic substances like iron, when

Magnetic induction.

subjected to a magnetic force, themselves become magnetized. Nickel and cobalt show similar properties, though to a much smaller degree. Certain alloys, too, of manganese, copper and aluminium are found to be highly magnetic, though made of non-magnetic elements. In all these cases, a north-seeking pole, presented to one end of a bar of the metal, induces a south-seeking pole on the nearer end. Attraction results, and the magnetized bar tends to move into the stronger parts of the field.

If we examine a bar of bismuth in the same way, using a very powerful electromagnet to produce an intense magnetic force, these phenomena are reversed. A pole of the same name as the inducing pole appears on the nearer end of the bismuth bar, and repulsion follows. Bismuth always tends to move into the weaker parts of the field. These phenomena are dealt with at greater length on p. 88.

By the use of the intense magnetic fields produced by powerful electromagnets, all bodies may be shown to possess some slight trace of magnetic properties. It is remarkable, however, that iron shows these phenomena to a very much greater degree than any other material known; unless an exception be made at some future time, in favour of certain rare metals such as erbium and others, which, should they be obtained in a pure form, seem likely to equal or exceed the susceptibility of iron.

Conditions of temperature exercise a determining influence in all cases of substances which respond to magnetic forces. Iron at ordinary temperatures is in many respects a different substance from the non-magnetizable metal into which it is transformed above its critical temperature; its magnetization, however intense, almost entirely disappears within a range of a few degrees' increase of temperature, and very marked alterations then take place also in many of its other properties, such as its specific heat and electrical conductivity. Nickel and cobalt are also affected by temperature in varying degrees; and certain alloys, notably those of iron and nickel, exhibiting two critical temperatures, have shown such great changes in magnetic properties with a change of temperature that they may be said to exist in two conditions, one of which is magnetic and the other not. In short, the magnetic properties of any substance depend on the quality of the substance itself, the amount of magnetization, on the temperature, and, as we shall see later, on the previous history of the substance.

The magnetization induced in a given bar of iron is evidently measured by the resulting magnetic moment; but it is convenient to have a unit which depends on the material only, and not on the shape or dimensions of a particular bar. Hence we form the conception of *intensity of magnetization*, and define it as the magnetic moment per unit volume. The ratio of the intensity of magnetization to the magnetic force producing it is called the *magnetic susceptibility*. It should be said that the magnetic force of this definition is the magnetic force actually effective within the substance of the iron. Owing to the magnetism induced at the ends of a piece of iron, the effective magnetic force is less than that applied externally, and depends on the shape of the iron. This result is considered below in § 33. The magnetic susceptibility is not a constant, but depends upon the value of the magnetizing force.

The magnetic force is defined as the mechanical force on a unit north-seeking magnetic pole, and, as long as the force is measured in air or other non-magnetic substance, this definition is free from ambiguity. When, however, the force is to be measured in iron, it is necessary to imagine that we cut a cavity in which to insert the measuring pole, and the resultant force on the pole will depend on the shape of the cavity. Where the lines of force leave iron and enter the cavity, a north-seeking pole is induced, and where they re-enter the iron, a south-seeking pole. If the cavity be a cylinder with its axis along the lines of force and its breadth very small compared with its length (Fig. 41), the effect of the ends is negligible, and the mechanical force on a unit pole within the cavity is still defined as the magnetic force.

Fig. 41.    Fig. 42.

If we take the other extreme case, and imagine the cavity in the form of a narrow thin crevasse, with its plane normal to the lines of force (Fig. 42), the effect of the induced poles on the opposite faces of the crevasse becomes very marked. The intensity $I$ of magnetization is defined as the magnetic moment per unit volume of the iron. If we consider a certain volume of the iron in the form of a rectangular block with one face on the side of the crevasse, we have

$$I = \frac{M}{La},$$

where $L$ is the length of the block, $M$ its magnetic moment, and $a$ the area of its face on the crevasse. But the magnetic moment of the block is defined as the strength of one of its poles multiplied by its length. Thus

$$I = \frac{mL}{La} = \frac{m}{a},$$

a result which shows that, in such a case, where the magnetism of the pole may be regarded as uniformly distributed over the end of the iron, the intensity of magnetization is equal to the strength of pole per unit area, that is, to the surface density of the magnetism.

The analogy between the crevasse in the iron and a system of two parallel metallic plates charged with opposite kinds of electricity to a surface density $\sigma$, will now be apparent, and by § 17 we see that the magnetic force between the two faces, due to the magnetism on them, is $4\pi I$. The total force on a unit pole in the crevasse will be the sum of this quantity and the magnetic force $H$ in a long cylinder, and this total force is defined as the *magnetic induction*, and usually written as $B$. If the direction of the magnetic induction coincides with that of the magnetic force, as we have supposed, we get the relation

$$B = H + 4\pi I.$$

We have already defined the quantity known as the magnetic susceptibility $k$ as the ratio $I/H$. Thus

$$B = H + 4\pi H k$$

$$= H(1 + 4\pi k).$$

The ratio of the magnetic induction in the iron to the magnetic force is a useful quantity, and is called the *magnetic permeability*, $\mu$. Therefore $\mu$ is $B/H$, and we get

$$\mu = 1 + 4\pi k$$

as the relation between the permeability and the susceptibility.

We have considered already the general form of the lines of magnetic force, and traced their analogy with the corresponding lines of electric force. On page 54 we developed the idea of lines of force into that of tubes of force, and agreed to imagine one tube of force as springing from one unit of positive electricity and ending on one negative unit.

With regard to lines of magnetic force, a different convention is adopted. Instead of drawing one line or tube of magnetic force from each unit of magnetism, we draw through unit area in the field, normal to the direction of the magnetic force, a number of lines of force equal to the numerical value of the magnetic force at that position. Throughout regions of the field in which there are no magnetic substances, the lines of force will be continuous. When magnets, or magnetic substances, are present, further consideration is necessary.

The laws of magnetic force are similar to those of electric force, and Gauss' theorem, § 12, and the deductions from it, hold equally in the case of magnetism. Moreover, as we shall see presently, the magnetic permeability $\mu$ corresponds with the dielectric constant $k$. Let us draw round an isolated magnetic pole an imaginary sphere. By analogy with the electric case considered in § 12, we then have the relation

$$\Sigma \alpha H \mu = 4\pi m,$$

where $H$ is the normal magnetic force on each element of area of the sphere, and $m$ the quantity of magnetism within the sphere, that is, the strength of the magnetic pole.

Now the quantity $H\mu$ is, by definition, equal to $B$ the magnetic induction in the medium, and hence $\Sigma \alpha H \mu$ is $\Sigma \alpha B$, a quantity which represents what we call the total magnetic induction or the magnetic flux through the surface of the sphere.

If we imagine the magnetic pole to be placed in air, and surrounded by a hollow spherical shell of iron, we could still apply Gauss' relation to the external surface of the shell in the form

$$\Sigma \alpha B = 4\pi m$$

Drawing tubes of induction from the pole to inclose proper elements of area, we see that the quantity $B$, the induction, is continuous throughout such a tube whether it passes through one medium or through more than one. This shows the importance of the conception of magnetic induction : it is the vector quantity which is continuous throughout any magnetic field.

If we draw through unit cross area surrounding a point a number of tubes of induction equal to the value of the magnetic induction at the point, in accordance with the usual convention, it should be noticed that the number of tubes through an area $a$ is $aB$. Hence, the total number of tubes proceeding from a pole of strength $m$ is $4\pi m$, and, from a pole of unit strength, the number is $4\pi$. This result should be compared with the electrical convention, by which one tube of electrostatic induction is imagined to spring from unit charge of electricity.

These relations may be considered profitably from another point of view, by supposing a narrow gap cut in an iron circuit otherwise continuous. Let us take, for instance, the case of a ring of iron, round which a continuous magnetizing force is exerted by the action of an electric current circulating in coils wound over the ring. Across a thin cut in the iron, the lines of force will run from one side to the other, and, if the cut be indefinitely thin, an indefinitely small number of lines of force will escape at the edges of the cut[1]. All the lines may then be imagined to run straight from one side to the other. The

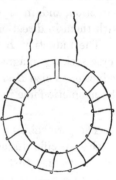

Fig. 43.

magnetic induction in the cut, which acts as the ideal crevasse of p. 76, is

$$B = \mu H = H\,(1 + 4\pi k).$$

But, as we have seen above, of this induction, $H$ is due to the magnetizing force, and $4\pi k H$, or $4\pi I$, is due to the induced magnetism on the faces of the cut. But $I$ is measured by the quantity of magnetism $m$ per unit area, and thus from each unit of area

---

[1] In practice there would always be considerable leakage of lines, for a cut of any finite thickness would throw back the poles some little distance from the faces of the cut. This, however, does not affect the reasoning given in the text, which is based on ideal conditions.

the amount of induction is $4\pi m$. Now, as stated, the conventional mode of drawing magnetic lines of force is to draw through unit area, normal to their direction in the field, a number of lines equal to the force at that place. Thus, in the case now under consideration, from each unit of area of the cut, $4\pi m$ lines proceed. From a unit pole, then, the number of magnetic lines of force is $4\pi$, in accordance with the result already obtained.

The lines crossing unit area of such an ideal cut represent the magnetic induction across the cut, and, also by what was said on page 76, the induction in the iron across any section of the ring, for it is possible to imagine an indefinitely thin crevasse cut anywhere round it. Thus continuous lines of magnetic induction may be imagined to pass round the ring, and across the intervening gap. In air, these lines are also lines of force, but in iron the lines of force would be $\mu$ times less in number than the lines of induction, and, in crystalline media, need not necessarily coincide with them in direction.

The induction $B$ in the air gap is the force acting on a unit north-seeking magnetic pole placed therein. In the iron the induction is also $B$, since it is continuous. But $B$ is $\mu H$, and thus the magnetic force in the iron is $1/\mu$ of its value in air.

The permeability $\mu$ in magnetism is now seen to be analogous to the specific inductive capacity or dielectric constant $k$ in electrostatics. Just as the electric force in paraffin is $1/k$ of its value in air, so the magnetic force in iron is $1/\mu$ of its value in air, and the complete expression for the mechanical force between two magnetic poles separated by a distance $r$ is

$$F = \frac{m_1 m_2}{\mu r^2}.$$

While, however, the dielectric constant is not found to vary when the electric force is changed, the permeability, as we shall see, depends on the applied magnetic force.

Another analogy, between the magnetic permeability of iron and the specific electric conductivity of an electric conductor, is useful in the theory of electromagnetic machinery, and will be considered in § 57, Chap. VII.

The conception of lines of magnetic induction is of great importance in electromagnetic theory, and will be used largely in Chapter VII. It should be noted that while electric or magnetic lines of force begin and end on electric charges or magnetic poles, lines of magnetic induction have neither beginning nor end, but form continuous closed curves.

The induction, by one pole of a permanent magnet, of an opposite pole on a piece of iron in its neighbourhood will cause the lines of force from the one pole to concentrate on the other, just as the lines of force from the north-seeking pole of a magnet in Figure 27 on page 51 concentrate on the south-seeking pole. It follows that a piece of soft iron, placed in a uniform magnetic field, deflects the lines of induction so that they crowd through

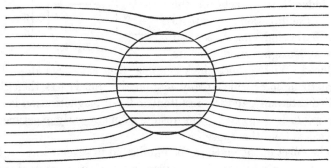

Fig. 44.

the iron. For certain cases it is possible to calculate this effect, and Figure 44, derived from Lord Kelvin's Reprinted Papers, shows the distribution of lines of electric force in the analogous system of a sphere of some dielectric substance, with a specific inductive capacity greater than unity, placed in air in a uniform electric field. The figure represents equally well the magnetic lines of induction through a paramagnetic substance such as iron, placed in a uniform magnetic field. When the permeability is very great, the magnetic induction inside the sphere is $3H$, where $H$ is the magnetic force in the undisturbed uniform field outside, while the magnetic force in the iron is $3H/\mu$—a value very small compared with that without.

Owing to this concentration of the lines of induction in ferro-magnetic substances, thick iron screens may be used to protect galvanometers or other magnetic instruments from the influence of an external field. The shielding effect is much less complete than the electrostatic shielding of conducting screens (p. 20), for an electric conductor is analogous to a substance of very great magnetic permeability.

**33.** In order to examine the magnetic properties of a sample of iron, certain precautions are necessary. When a bar

Experiments on Magnetic Induction.

of iron is placed in a magnetic field, poles are induced near its ends, and these poles will themselves exert a magnetic force on the substance of the bar. It is evident from

Figure 45 that the force due to the induced poles acts in the iron in the direction opposite to that of the inducing force, and thus a demagnetizing effect is

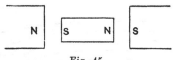

Fig. 45.

produced. With a short, thick bar this effect is considerable. The magnetizing force which really acts on the iron is less than that applied externally, and the resultant induced magnetization is less than would be expected. The results of many early experiments are useless from this cause.

The effect of the ends can be eliminated in several ways. We may, if we please, use an endless ring of iron, and observe its influence on the electromagnetic induction of currents in coils wound over it by methods that will be described in Chapter VII.

We may, however, in certain cases, use iron rods, apply a constant magnetizing force, and measure the resultant magnetiza-tion by the usual deflection methods. If a bar in the form of an elongated ellipsoid be used, it is possible to calculate the reverse effect, or, by taking a long, thin wire, with a length two or three hundred times the diameter, the effect of the ends is so small that, in ordinary work, it may be neglected.

We shall show in the next chapter that inside a long helical coil of wire, called a solenoid, the magnetic field of force is uniform, and the magnetic force equal to $4\pi nc$, where $n$ is the number of turns of wire per unit length, and $c$ is the electric current flowing in the coil measured in electromagnetic units.

An iron wire, or bundle of iron wires, placed inside a solenoid so that the coil well covers the ends, will therefore be subject to the Earth's field and to a nearly uniform magnetic force of known and controllable strength.

The solenoid is fixed horizontally, east and west. At a point on its axis produced is placed a magnetometer consisting of a suspended needle and mirror (Fig. 34, p. 67). Another arrangement is to fix the solenoid vertically, and place the magnetometer needle on a level with its upper pole.

The current is passed in series through the solenoid, and through a small coil which is moved about till, whatever current be passed, it just balances the effect of the empty solenoid on the magnetometer needle. The iron is then inserted in the solenoid, and a small current passed round the coils. The deflection of the needle is due then to the iron alone. Assuming that the poles are very near the ends of the long wire, the strength of the magnetic poles may be calculated from the deflection by the law of inverse squares. From the pole strength the value of the magnetic moment is deduced, and this quantity, divided by the volume of the iron, gives the intensity $I$ of magnetization. The susceptibility $k$ of the iron is the ratio of the intensity of the magnetization to the magnetizing force $H$, the permeability $\mu$ is $1 + 4\pi k$, while the magnetic induction $B$ is the product of $\mu$ and $H$.

Another method, already referred to, which is nearly always used in practice, depends on the induction of a secondary electric current in one coil of wire by a variation of a primary current in another coil in the neighbourhood. If both coils are wound over an iron core, the induced electromotive force, being proportional to the rate of change of the magnetic induction, gives a means of measuring the properties of the iron. The electrical principles used in this method are explained in § 53, Chapter VII.

It is not convenient to cut the sample to be tested in the form of a ring, and an arrangement known as the Epstein square is better (Fig. 46). Test strips of the iron weighing about 4 lbs. are built up into bundles, the strips being insulated from each other to maintain a uniform flux. The bundles are placed in four solenoids placed at the corners of a square, the magnetic circuit being completed at the corners by small angle pieces of the test material clamped to the strips.

Each solenoid has three windings, one primary and two secondaries By reversing an electric current in the primary circuit, an electromotive force depending on the magnetic flux in the iron is set up in the secondaries, and by connecting one of them through a ballistic galvanometer (see p. 113) the flux can be estimated. But a more accurate method is to pass an alternating current round the primary circuit. A volt-

Fig. 46.

meter, such as the electrostatic form described on p. 49, is connected across the terminals of one secondary to give the magnetic flux, and a wattmeter across those of the other to give the energy losses (see below).

The results of such experiments, by whichever method conducted, can be represented graphically by plotting abscissæ proportional to the magnetizing force $H$, and ordinates proportional either to the magnetic induction $B$, or to the intensity of magnetization $I$.

In starting the experiments, the iron may be demagnetized by heat, or by an alternating magnetic field, the intensity of which, large to begin with, is diminished gradually to the vanishing point.

Figure 46$a$ shows a curve, obtained by Sir J. A. Ewing, giving the magnetic induction. When the magnetic force is small, less than about one-tenth of the earth's horizontal force, the induction in demagnetized iron increases nearly in proportion to the force, as shown near $O$ in the figure. Then it rises much more quickly, but finally increases from $A$ onwards more and more slowly, till, at the highest forces possible, the induction is only increased very slowly by the further increase of magnetic force, as from $G$ to $P$ in the figure; here the intensity of magnetization becomes nearly constant, and the iron seems to approach a state of magnetic saturation.

The existence of this state of saturation has been confirmed by subjecting iron to very much stronger magnetic forces than can be obtained in coils carrying currents, where the field cannot much exceed 2000 gausses or C.G.S. units. By placing small iron bars in the concentrated fields obtained between the poles of powerful electro-magnets, Ewing and Low subjected them to fields approaching 20,000 units. The induction was measured by suddenly withdrawing the iron and observing the induced current in a coil of wire wound over it.

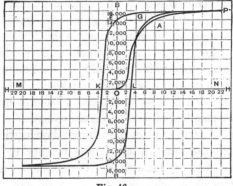

Fig. 46 a.

If, after magnetizing iron, the magnetic force be decreased steadily, and measurements taken for different forces, the curve of magnetization or induction does not return along its old course. The iron exhibits magnetic retentiveness, and, even when the magnetizing force is wholly removed, keeps a large proportion of its maximum magnetization. If the force be reversed, the induction falls rapidly, as shown by the part $EK$ of the curve. As the reversed force is increased, the iron becomes saturated with the magnetization in the other direction. On decreasing this force, retentiveness is again shown, and the reversed magnetization is only removed when the force is once more reversed; the curve then soon rises to $L$. A new cycle of magnetic force does not cause the magnetization to pass along the original path $OA$, but along the curve arising from $L$.

The general lagging of the induced magnetization behind the magnetizing force has been named by Ewing *hysteresis*. The particular curve shown in Figure 46 $a$ must be regarded merely as typical; the exact form depends on the sample of iron, and on the treatment to which it is subjected. Curves are sometimes drawn between the magnetic force $H$ and the intensity of magnetization $I$. Such curves may be obtained from the curves giving the magnetic induction $B$, as in Figure 46 $a$, by considering the relation

$$B = H + 4\pi I.$$

From the ordinates of Figure 46 $a$, the value of $H$ must be subtracted, and the result must be divided by $4\pi$, in order to get the curve of intensity of magnetization. It follows also that the area of the closed curve in the induction diagram of Figure 46 $a$ will be $4\pi$ times the area of the magnetization curve.

The areas of these curves have an important physical significance, which will become apparent if we consider the work done in putting a piece of iron round the cycle of changes represented in the diagram.

We may imagine that we magnetize a magnet by separating a series of very small equal and opposite quantities of magnetism at one end, and carrying one of the separated quantities along the magnet to form the other pole. If the magnet be indefinitely short compared with its breadth, or if we imagine it to be cut out of a solid block of material, the magnetic force $H$ between its poles will be uniform, and the work done in separating the one small quantity of magnetism $\delta m$ from the other over the length $2l$ of the magnet is $2lH\delta m$.

This infinitesimal shift of magnetism involves an infinitesimal change $\delta I$ in the intensity of magnetization. If $\alpha$ be the area of cross-section of the magnet, we have (p. 76)

$$\delta I = \delta m/\alpha.$$

Hence the work done during the process we have described is

$$2l\alpha H\delta I.$$

But $2l\alpha$ is the volume of the magnet. Hence, measured in ergs, the work done per unit volume is

$$W = H\delta I.$$

By a process similar to that used on page 43, it follows that, in a magnetization curve corresponding to the induction curve in Figure 46 $a$, while the magnetization of the iron passes along the curve from $O$ to $P$, the total work done, or $\Sigma H \delta I$, is represented by the area $OPB$; while, throughout a complete cycle of changes, the excess of the work done in magnetizing over the work returned while demagnetizing, for unit volume of the iron, is represented by the area of the hysteresis loop in a magnetization diagram drawn between $H$ and $I$. This work is dissipated in the iron and appears as heat.

The area of the hysteresis loops on the $B$ and $H$ curves is, as we have seen, $4\pi$ times the area of the corresponding loops on the $I$ and $H$ curves. Hence the work done in ergs is represented by $1/4\pi$ times the area of the $B$ and $H$ loops in Figure 46 $a$.

When iron is to be used in electromagnetic machinery, it is important that a knowledge of the magnetic properties of different samples should be obtained. For most purposes it is desirable that the permeability should be high and the hysteresis small.

Practical instruments have been devised, whereby the permeability may be estimated quickly by the tractive force exerted on a standard block of iron by the sample to be tested when forming the core of an electromagnet. The hysteresis effect may be determined by an instrument invented by Ewing, in which the lagging of the magnetization in the sample bar is made to exert a couple on a permanent magnet, between the poles of which the bar revolves.

The permeability ($\mu = B/H$) is not constant as the magnetic force changes, as is seen from the varying slope of the curve in different parts of the $BH$ diagram, or from the following table of experiments on strips of steel.

| $H$ | $B$ | $\mu$ |
|---|---|---|
| 1 | 1580 | 1580 |
| 2 | 4930 | 2465 |
| 3 | 7000 | 2333 |
| 10 | 11800 | 1180 |
| 20 | 13840 | 692 |
| 50 | 15800 | 316 |
| 100 | 16600 | 166 |
| 160 | 17980 | 112 |

The permeability is greatest for moderate fields, where the curve rises most rapidly, and falls off rapidly as the field increases and the iron approaches saturation. At ordinary temperatures, for good soft iron, it may vary from 2000 to 4000. For small values of $H$, it increases considerably with the temperature, till a critical temperature of 700° to 900° is reached, when the permeability falls rapidly to unity, and the susceptibility vanishes, iron becoming a non-magnetic substance till recooled. That this is connected with the internal structure of the iron is shown by the fact that, at the same critical temperature, the phenomenon known as recalescence appears. If a piece of iron be heated above this temperature till nearly white-hot and allowed to cool, it changes to a dull, almost invisible redness, and then brightly glows again as it passes the critical temperature. This behaviour probably means that a change in the crystalline structure of the iron occurs about the temperature in question—a change which involves latent heat. As the iron loses heat it may become undercooled, just as water kept without disturbance may be cooled below its freezing-point. When the structural change sets in, it will then proceed rapidly, and a large amount of latent heat will be evolved; thus the temperature may rise considerably. It is likely that some such structural change produces the marked alteration in magnetic properties which iron displays at the temperature of recalescence. The rise of temperature may be demonstrated by drawing a curve between temperature and time as a sample of iron is cooled.

While iron is the most magnetic substance known, all bodies, when placed in the strong magnetic field between the poles of a powerful electromagnet, show magnetic properties. With some substances, known as paramagnetic substances, the magnetic effects are the same in kind as in iron and other ferro-magnetic bodies though less in degree. With others, or diamagnetic substances, the effects are reversed, so that a bar of bismuth, for example, moves out of a magnetic field, and acquires poles of the opposite sign to those induced in iron.

The magnetic susceptibilities of some of the elements are given in the table on p. 89. Paramagnetic elements are marked + and diamagnetic elements −.

### Solids.

| | $k \times 10^7$ | | | $k \times 10^7$ |
|---|---|---|---|---|
| Aluminium | ... + 6·5 | | Phosphorus | ... − 9 |
| Antimony | ... − 9·5 | | Platinum ... | ... +13·2 |
| Bismuth ... | ... − 14 | | Potassium... | ... + 4 |
| Chromium | ... + 37 | | Silver | ... − 2 |
| Copper ... | ... − 0·87 | | Sulphur ... | ... − 5 |
| Manganese | ... +106 | | Zinc | ... − 1·5 |

### Liquids.

| | | | | |
|---|---|---|---|---|
| Bromine ... | ... − 4·1 | | Liquid Oxygen | ... + 3·24 |
| Mercury ... | ... − 1·9 | | Water | ... − 8 |

### Gases.

| | | | | |
|---|---|---|---|---|
| Helium ... | ... − 0·02 | | Nitrogen ... | ... + 0·24 |
| Hydrogen | ... − 0·08 | | Oxygen ... | ... + 1·23 |

These numbers are to be compared with those for the ferro-magnetic elements:

Iron 100 to 300, Nickel about 4, Cobalt about 6.

Thus in iron the susceptibility is some $10^8$ times as great as in para- and diamagnetic bodies.

A mixture of manganese, aluminium and copper (24 Mn, 16 Al, 60 Cu) known from its discoverer as a Heusler alloy, though made of elements only slightly magnetic, must be classed as a ferro-magnetic substance, its susceptibility being about 2. Its properties show that the strong magnetic effect known as ferro-magnetism must be an affair not of atoms, but of molecules or groups of molecules.

The explanation of the weaker effects known as para- and dia-magnetism will be dealt with below.

**34.** If a magnetized steel knitting-needle be broken in halves, each half will be found to act as a permanent magnet with poles near its ends. This process may be repeated as long as the fragments of iron are large enough to be broken—each fragment retains its magnetic properties. If the fragments were put together again, it is evident that the intermediate poles would neutralize each other, leaving the original

*Theories of magnetism.*

poles of the whole magnet effective at the ends of the chain. Such observations and considerations have suggested that magnetism may be an atomic or molecular property, the individual atoms or molecules, or groups of molecules, being imagined as diminutive magnets. The fact that alloys of the non-magnetic metals manganese, copper and aluminium show magnetic properties indicates that, in these cases at all events, magnetism depends on the molecules or groups of molecules rather than on the elementary atoms.

An explanation of the phenomena of gradual magnetization was given by Weber, who suggested that, in an unmagnetized iron bar, the axes of the individual molecular magnets lie in all sorts of irregular and arbitrary directions. As a magnetic force is applied, more and more of the molecules set so that their axes point in the same direction. Contiguous poles of the molecular magnets neutralize each other except at the ends of the bar, where the effective poles of the whole magnetic system appear.

That some rearrangement of the molecules is involved in the process of magnetization seems indicated by the changes in volume which then occur. Joule observed that the elongation of a bar of iron or soft steel on magnetization was, up to a certain point, proportional to the square of the intensity of the magnetizing force; but that it failed to comply with this relation some time before the saturation point was reached. It has since been shown that, if the magnetizing force be pushed beyond the strength with which Joule experimented, the extension of the bar ceases, and the bar gradually returns, first to its original length, and ultimately recedes within that limit. A cobalt bar, on the other hand, contracts in the early stages of magnetization and afterwards recovers its original length, and then increases rapidly with increasing intensity. A bar of nickel appears to diminish in length throughout the whole process of magnetization.

In order to explain the fact that the state of magnetic saturation is not reached on the application of a small magnetic force, Weber supposed that the motion of the molecular magnets was opposed by a frictional resistance. But this does not explain the phenomena of residual magnetization, and Ewing showed that the supposition of friction was unnecessary as well as insufficient. Ewing suggested that a collection of little magnets, turning without friction, would, even in the absence of a magnetizing force, set themselves in stable groups under the influence of their own mutual forces. The groups will have all sorts of configurations throughout the volume of the iron ; the bar will possess therefore no resultant magnetic moment. On the application of a magnetic force of gradually increasing intensity, the first effect is to deflect slightly some of the magnets from their original positions. This deflection will, for small angles, be proportional to the magnetic force, and, if the magnetic force be removed, the molecules will revert to their original state. These relations explain the first stage in the actual magnetization of iron—the first part of the curve in the neighbourhood of $O$, in Fig. 46 $a$.

As the magnetic force is increased, some of the groups of molecular magnets become unstable and break up. More and more of the little magnets set in the direction of the impressed force. When nearly all of them are so arranged, the iron approaches its state of saturation.

Ewing constructed a model, consisting of a number of small compass-needles pivoted and placed on a board. The board was fixed within a solenoidal coil of wire, through which an electric current could be passed. In this way a hysteresis curve was obtained—a curve which reproduced in a very striking manner the phenomena of the magnetization of iron.

Nevertheless this early model fails to represent completely the phenomena of ferromagnetism, and in 1922 Ewing described a new model in which the magnetic element is a pivoted needle placed inside a framework of other magnets. On this view the real magnetic element is some inner parts of the atom resting in one of several possible positions of stability under the influence of the outer atomic shell.

The structure of the individual magnetic elements remains to be considered. As we shall see in the next chapter, an electric current flowing in a closed circuit acts as a magnet. Long ago Ampère suggested that the conception of electric currents flowing in minute circuits, round or within the individual molecules, might furnish a more fundamental explanation of magnetic phenomena. And, of recent years, this hypothesis has gained added significance. As we shall see in the sequel, there is reason to believe that the

properties of atoms may be explained by the supposition of negatively electrified corpuscles, which constitute isolated electric
charges, revolving in orbits round a positive nucleus. It has been
shown experimentally, by Rowland and others, that a moving
charge of electricity is equivalent to an electric current. The
requirements of Ampère's hypothesis are thus satisfied by the
latest theory of the atom, though why iron should so far transcend
all other known substances in magnetic properties remains a
problem for future elucidation.

The susceptibility of diamagnetic substances, except in bismuth
and antimony, is found to be independent of temperature. This
suggests that it is an atomic property, and an explanation of it can
be given by considering the effect of a magnetic field on the orbits
of the atomic electrons. The effect will be to set up an opposing
electromotive force and thus to decrease the resultant magnetic
induction.

On the other hand Curie has found that, for paramagnetic
substances, the susceptibility referred to unit mass varies inversely
as the absolute temperature. This indicates that in paramagnetic
bodies the atoms or molecules already possess magnetic moment,
probably produced by the revolution of an electron uncompensated
by another, and that the effect of a magnetic field is to rotate the
atom or molecule into a new orientation and thus increase the
total magnetic moment.

Ferromagnetism is explained if we suppose that a number of
paramagnetic molecules have enough influence on each other to
form larger molecular groups themselves influenced by the magnetic field, perhaps as in Ewing's model.

## REFERENCES.

*Magnetic Induction in Iron and other Metals*; by J. A. Ewing.
Articles on Magnetic subjects in the *Dictionary of Applied Physics*.

# CHAPTER V

## THE ELECTRIC CURRENT

Discovery of the electric current. Voltaic cells. Effects of the galvanic current. Relations between currents and magnets. Magnetic force inside a solenoid. Magnetic force due to a long, straight current. Ampère's hypothesis. A circular current. The tangent and ballistic galvanometers. Electromotive force. Ohm's Law. Electric Resistance. Wheatstone's bridge. Specific resistance. Comparison of electromotive forces. Heating effect of an electric current.

**35.** THE different forms of apparatus for the production of electricity, hitherto discussed, are all intended primarily to enable us to give a static charge of electricity to some insulated conducting body. It is true that, if a conducting circuit be formed, connecting the collecting apparatus of an electric machine with the rubber or with the earth, a more or less continuous flow of electricity must proceed along the circuit. Even in the most elaborate form of influence machine, however, the amount of electricity passing in a second is so small that very sensitive arrangements are necessary to detect the current in the conducting wires; though, if an air gap be interposed, the high differences of electric potential produced by the machine result in visible sparks and the attendant phenomena.

At the beginning of the nineteenth century, a new field of research was opened up by the discovery of the galvanic or voltaic cell. This arrangement gave rise to a series of phenomena grouped originally under the name of galvanism, which, by the efforts of many observers, was gradually brought into relation with the older electricity. Faraday may be said to have established finally the identity of the two manifestations. He showed that a galvanic current was nothing more nor less than a flow of electricity, enormous in quantity compared with that given by an electric machine, but driven along by potential differences very much less than those involved in the older type of apparatus. Since no accumulation of electricity can be detected at any point in the circuit, it follows that the current may be represented figuratively by the flow of an incompressible fluid along rigid and inextensible pipes. We define the strength of the electric current, as we shall see later, by means of the magnetic force it produces, and therefore we must not define it by means of the quantity of electricity passing. We must reverse this procedure, and make a new definition of quantity of electricity. Unit quantity of electricity passes any cross-section of a circuit when a current of unit strength flows round that circuit for unit time. The relation of this definition of unit quantity of electricity with the statical definition based on the mechanical forces between charged bodies will be considered in a future chapter.

The discovery of the voltaic cell was due to a chance observation, which seemed at first to lead in a different direction—an experience not uncommon in the history of scientific investigation. About the year 1786, an Italian named Galvani noticed that the leg of a frog contracted under the influence of a discharge from an electric machine. Following up this discovery, he observed the same contraction when a nerve and a muscle were connected with two dissimilar metals, placed in contact with each other. Galvani attributed these effects to a so-called animal electricity, and it was left for another Italian—Volta, of Pavia—to show that the essential phenomena did not depend on the presence of an animal substance. In 1800 Volta invented the pile known by his name, which, in the opening years of the following century, provided a means of investigation yielding results of intense interest in the hands of the discoverer and his contemporary workers in other countries. The scientific journals of the time are full of the marvels of the new science, the study of which was taken up with an ardour little short of that shown a century later in the elucidation of the phenomena of radio-activity.

**36.** Volta's pile consisted of a series of little discs of zinc,
Voltaic cells.     copper, and paper moistened with water or brine,
placed one on top of the other in the order—zinc, copper, paper, zinc, etc....finishing with copper. Such an arrange-

ment is really a primitive primary battery, each little pair of discs separated by moistened paper acting as a cell, and giving a certain difference of electric potential, the differences due to each little cell being added together and producing a considerable difference of potential (or electromotive force as it is now called) between the zinc and copper terminals of the pile. Another arrangement was the crown of cups, consisting of a series of vessels filled with brine or dilute acid, each of which contained a plate of zinc and a plate of copper. The zinc of one cell was fastened to the copper of the next, and so on, an isolated zinc and copper plate in the first and last cell respectively forming the terminals of the battery. Volta thought that the origin of the effects was to be sought at the junctions of the two metals; hence the order of the discs in the pile and the terminal metal plates in air in the crown of cups. These plates, and the corresponding discs in the pile, were soon found to be useless, though they figure extensively in early pictures of the apparatus.

If a current be taken from Volta's pile or crown of cups by connecting the terminals by means of a wire, that current diminishes rapidly in intensity; this is due chiefly to a film of hydrogen which forms on the surface of the copper plate. In a later chapter we shall study the theory of such phenomena under the head of electrolytic polarization. Here we are concerned merely with the practical means adopted to eliminate its effects in voltaic cells. Cells can be classed in three groups, according as the depolarizing action is mechanical, chemical, or electrochemical. The following examples may be given:

*Cells mechanically depolarized.* Smee's cell, where a silver plate is covered with crystals of platinum, the sharp edges of which aid the escape of the hydrogen.

*Cells chemically depolarized.* 1. The bichromate battery. The zinc and carbon plates are surrounded with an oxidizing mixture of sulphuric acid and a strong solution of potassium bichromate. The hydrogen, instead of being evolved as gas, is oxidized to water.

2. The Leclanché cell, consisting of zinc in a solution of sal-ammoniac, and carbon surrounded with a mixture of broken carbon and manganese dioxide.

3. Grove's cell, where one plate is zinc in dilute sulphuric acid, and the other is platinum in strong nitric acid contained in a porous pot.

*Cells electrochemically depolarized.* 1. Daniell's cell, in which zinc is placed in dilute sulphuric acid or a solution of zinc sulphate, and a porous pot contains copper immersed in a strong solution of copper sulphate. Copper is deposited on the copper plate instead of hydrogen.

2. Latimer Clark's cell, used as a standard of electromotive force. Here a zinc rod is placed in a solution of zinc sulphate; and mercury is covered with a paste of mercurous sulphate, which deposits mercury when a current flows.

**37.** As we have said, the discovery of the galvanic current was made by the detection of its physiological effects on the leg of a frog. As soon as the invention of the voltaic cell placed a more powerful instrument in the hands of investigators, it was found that striking chemical changes accompanied the flow of the current through water and aqueous solutions. To these phenomena the early experimenters chiefly directed their attention. A detailed study of this important branch of our science will be found in the chapter on electrolysis, and we shall now pass to other subjects.

*Effects of the galvanic current.*

It was soon found that, when passing through a conductor of any kind, the current evolved heat, the amount of which depended on the nature of the conductor. This thermal effect also will be considered later. In this place we shall depart from the chronological order of development, and pass to another property of the current, namely, its power of deflecting a magnetic needle. This power was discovered by Oersted of Copenhagen in 1820. Its importance for our present purpose lies in the fact that, by the magnetic force which a given current will produce, the strength or intensity of that current is, by general agreement, defined and measured. Moreover, the magnetic effects of minute currents give the most sensitive means of detecting them.

By a convention universally adopted, we agree to suppose that an electric current flows in the connecting wire from the positively electrified copper (or carbon) to the negatively electrified zinc plate, and within the battery from the zinc to the copper (or carbon) plate. In accordance with this convention, the copper plate is called the positive, and the zinc plate the negative terminal of the battery.

If a wire, along which a current is passing from south to north, be placed over and parallel to a compass-needle, the north-seeking pole is deflected towards the west. If the wire be placed below the needle, with the current still passing from south to north, this deflection is reversed. Thus, if the wire be wound in a coil, so that the current passes in one direction above the needle and returns in the other direction below, the effects on the needle are of the same sign, and the deflection is multiplied greatly. Such an arrangement is known as a galvanometer. Except in special cases, the chief use of a galvanometer is to detect the presence and direction of electric currents; hence the object in designing a galvanometer is to increase the sensitiveness; it is not usually necessary to know the relation between the strength of the current and the deflection of the needle.

A very sensitive galvanometer, invented by Lord Kelvin, is shown in Figure 47. It consists of a coil of wire closely surrounding a small suspended mirror, on the back of which are fixed several pieces of magnetized watch-spring, as in the magnetometer illustrated in Figure 34 on page 67. As we have seen, the deflection of a needle produced by a magnetic force $F$, applied at right angles to the original position of the needle, is given by

$$F = H \tan \delta,$$

where $H$ is the strength of the field (due to the earth or other magnetic system) in which the needle hangs. The deflection produced

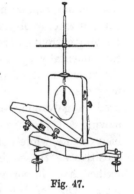

Fig. 47.

by a given force, that is, by a given current, will thus be inversely proportional to $H$, the strength of field. The sensitiveness of the galvanometer therefore will be much increased by the use of a control steel magnet, which can be moved into such a position that it nearly counteracts the earth's horizontal magnetic force.

The sensitiveness has been increased still further by fixing to a light rigid glass fibre two systems of light steel strips with their poles in opposite directions. The magnetic moments of these magnets are adjusted carefully till they are nearly equal as well as opposite, and the astatic system is then suspended by a quartz fibre so that each half lies within a coil. The current flows in opposite directions in the two coils, and the deflection is read with a telescope and mirror. A current of $10^{-10}$ ampere (see page 118) may thus be detected.

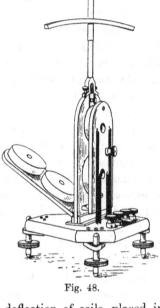

Since a current produces a magnetic force, it will itself experience a force when placed in a magnetic field, whether that field is due to permanent magnets or to other currents. The phenomena of attraction and repulsion between circuits of wire, freely suspended and carrying currents, were investigated

Fig. 48.

experimentally by Ampère, and the deflection of coils, placed in the strong field of permanent magnets, is now extensively used in galvanometers of the moving coil type (Fig. 49). Such instruments, while not quite so sensitive as those already described, have important advantages. One of the chief advantages of these galvanometers is their freedom from disturbance when the external field of magnetic force is liable to some amount of variation, owing to

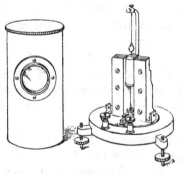

Fig. 49.

the movement of masses of iron, etc. The field produced by the

permanent magnets is so very strong, compared with that of
the earth and other external objects, that a small variation in the
external field is inappreciable. In the d'Arsonval galvanometer,
the current is led into the coil by a phosphor-bronze strip, which
also acts as a means of suspension, and is taken out by a torsionless
coil of fine silver wire.

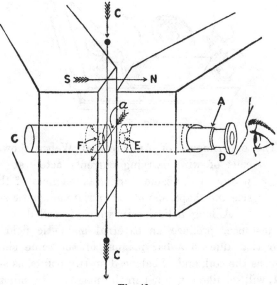

Fig. 49 a.

In the Einthoven galvanometer the moving coil type of instrument is reduced to its simplest form. A fine wire or silvered quartz
fibre is stretched in a strong magnetic field. When a current passes
through the wire it is displaced across the field, and the displacement measured with a microscope (Fig. 49 a).

In practical ammeters and voltmeters, the moving coil is
mounted on pivots between jewels, and carries a pointer which
moves over a scale divided so as to read amperes or volts directly
(Fig. 50). These instruments are usually made with a very high
resistance, so that the current through them is directly proportional
to the voltage applied. On the other hand, a heavy current may
be measured by passing it through a known resistance and using
the instrument to give the voltage between the ends of that

7—2

resistance (see p. 133). The same instrument can thus be used either as an ammeter or a voltmeter.

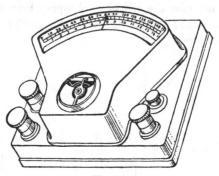

Fig. 50.

**38.** The experiments of Ampère and Weber showed that coils of wire carrying currents acted at external points in the same manner as magnets of the same size and shape, and of appropriate magnetic strength.

Relations between currents and magnets.

A long helical coil, which is called a solenoid, will, for instance, produce an external magnetic field exactly similar to that due to a bar magnet of the same shape and dimensions as the coil, and, if balanced on two points, as shown in Figure 51, will set like a magnet in the magnetic meridian of the earth. As the length of the solenoid is decreased, the equivalent magnet becomes shorter also, and, if the coil be imagined as reduced to a single circle of wire, the equivalent magnet must be represented by a circular disc of steel, magnetized in a direction at right angles to its plane, so that one face of the disc is a large flat north-seeking pole, and the other face a similar south-seeking pole. Such a disc is known as a *magnetic shell*. A magnetic shell may be imagined as made up of a number of very minute bar-magnets placed side by side, with all the poles of one name pointing the same way.

If a small circuit be equivalent to a corresponding magnetic shell, it must follow that the same relation holds for circuits and shells of any size. A large shell may be imagined to be resolved into a number of small shells of equal strength. Each of these may be replaced by its equivalent current flowing round the edge of the little shell; and, since the currents too must be of equal strength, the effects of the currents along each internal line of junction will cancel, and we are left with one continuous current flowing round the external edge of the large shell.

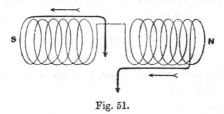

Fig. 51.

It will be evident that a closed electric current of any size and form is equivalent to a magnetic shell with its edge coinciding with the wire carrying the current. This equivalence between magnets and currents extends only to the external magnetic fields which they produce. Inside the coils and magnets the conditions are different.

As we have seen in § 28, the effect of a magnet at distant points depends on a quantity known as the magnetic moment. In the case of magnetic shells, it is convenient to use a corresponding quantity which does not depend on the size or shape of the shell. Hence we define the *strength of a magnetic shell* as the magnetic moment per unit area. From the experimental equivalence between currents and magnetic shells, it follows that the electromagnetic unit of current may be defined as that current which is equivalent to a magnetic shell of unit strength when both are in air.

If two circular currents be placed parallel to each other, it is clear that, if the currents circulate in the same direction in each circuit, the equivalent magnetic shells will have poles of opposite name facing each other. Thence follows a result readily verified by experiment, namely, that currents flowing in the same direction attract each other, while currents flowing in opposite directions repel each other.

In § 30, we found that the magnetic potential of a short bar-magnet at a point at a distance $r$ from the centre of the magnet was $M \cos \theta / r^2$, where $\theta$ is the angle between the magnetic axis of the magnet and the line joining its centre to the point considered. If we imagine the magnet shortened till it becomes vanishingly short, we get a small magnetic shell, and the same expression still gives the potential at an external point.

We defined the strength $S$ of a shell as the magnetic moment per unit area. Thus $S$ is $M/\alpha$, and the potential $V$ is given by

$$V = S \frac{\alpha \cos \theta}{r^2}.$$

Now $\alpha \cos \theta / r^2$ measures the solid angle $\omega$ which the shell subtends at the point $P$. Hence

$$V = S\omega.$$

For a large shell, the potential is the sum of the potentials due to the small shells of which we may suppose it to be constituted. Thus, if $\Omega$ be the solid angle subtended by a magnetic shell of any size at an external point $P$, the potential at that point is equal to the strength of the shell multiplied by the solid angle which it subtends at the point, for the strength of the shell is uniform all over its surface, and

$$V = \Sigma S\omega = S\Sigma\omega = S\Omega.$$

Since currents are equivalent to magnetic shells, a similar expression gives the magnetic potential at a point due to a current $c$. If $\Omega$ denote the solid angle subtended at the point by the circuit in which the current flows,

$$V = c\Omega,$$

unit current, in accordance with our definition, being taken as equivalent to a shell of unit strength placed in air.

Let us imagine a circuit of wire of any form through which passes a current $c$. Its magnetic effect on points outside it will be the same as that of an equivalent magnetic shell. At a point indefinitely near the plane of the shell, the solid angle which the shell subtends is half that of the whole surrounding sphere, or $2\pi$. The potential at that point due to the equivalent current is therefore $2\pi c$. At a corresponding point on the opposite side of the plane of the shell or current the solid angle is $-2\pi$, and the potential is $-2\pi c$. Thus the difference of potential between these two points is $4\pi c$.

The difference of magnetic potential between two points has been defined as the work done against the magnetic forces in carrying a unit north-seeking magnetic pole from one to the other. Thus, if we take a unit north-seeking magnetic pole from a point just outside the plane of a shell or current to a corresponding point on the other side, the path being in air round the edge of the shell or current, the work done is $4\pi c$ ergs.

To pass through the shell, and regain the original point, we must suppose the pole taken up through a hole in the shell. Here the work is clearly reversed, and, when we regain the original point, the total work done vanishes, as, indeed, follows from the principle of the conservation of energy. But here appears a difference between currents and shells. In completing the path round the current, there is no steel to pass. The path is completed through air, and there is clearly no discontinuity of force. Thus the work done in taking a unit north-seeking magnetic pole completely round from one side of the current

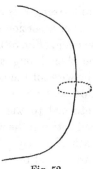

Fig. 52.

to the other is not appreciably added to by taking it through the indefinitely short path required to complete the circuit and return to the original point: the total work is still $4\pi c$.

This difference between magnets and currents is significant. It depends on the fact that in the current we have a source of energy maintained elsewhere, from which we may draw when moving a magnetic pole in the neighbourhood. In a magnet we have no such source of energy, and no work can be gained or lost by moving a pole, so that in the end it returns to its starting-point. The equivalence between shells and currents does not hold when we deal with points within them—it applies to external points alone.

In carrying a unit magnetic pole once round a current, the work done is independent of the nature of the medium through which the path lies. For that part of the forces on the pole due to the magnetization induced or permanent in the medium can be represented as due to a system of magnets, and, therefore, the work done by these forces round a closed circuit must vanish.

Thus, whenever a unit north-seeking magnetic pole is made to circulate once round a current $c$, work to the amount of $4\pi c$ is done whatever be the nature of the medium, and this result gives us a new definition of current strength.

**39.** An application of this result enables us to deduce the value of the magnetic field inside a solenoid, that is, a helical coil of wire through which a current passes.

*Magnetic force inside a solenoid.*

Let us imagine that an isolated magnetic pole of unit strength is carried along a path $AB$ inside and parallel to the axis of a portion of a very long solenoid (Fig. 53). Let the pole be then brought out between the wires of the coil, taken outside along the path $DE$, and returned to the inside of the coil between the wires along the

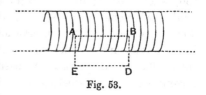

Fig. 53.

path $EA$. By § 38, the work done is $4\pi c$ for each turn of wire surrounded by the path, and, if $n_1$ be the number of turns of wire per unit length of the coil, the total work done is $4\pi n_1 c . \overline{AB}$.

The lines of force of the solenoid will be similar to those of a bar-magnet, except that, instead of ending on the poles, they form continuous curves by running back along the inside of the coil. Thus, inside the coil, all the lines of force are crowded into a small space, but outside, the same number are spread throughout the whole field. If the coil be very long, then the magnetic force near $DE$ is vanishingly small compared with the magnetic force inside the coil. Along $DE$ the force is negligible, and the work done when the pole is taken along it is negligible also. The portions $BD$ and $EA$ of the paths run at right angles to the lines of force, and, along them also, the force vanishes, and no work is done. The whole of the work throughout the path is concentrated into the length $AB$, and, throughout that length, the magnetic force is, by symmetry, uniform when the windings are very close together. Writing this magnetic force as $H$, the work is $H . \overline{AB}$ and

$$H . \overline{AB} + 4\pi n_1 c . \overline{AB}.$$

Thus                    $H = 4\pi n_1 c.$

Nothing has been said about the distance of the line $AB$ from the axis of the coil, and the result holds for all places inside the solenoid. The field of force, then, is uniform everywhere inside the solenoid, and acts in the direction parallel to the axis.

This result has been used already in describing the experimental method of determining the magnetization of iron, and is of great importance in the theory and manufacture of electromagnetic machinery.

The same result may be reached by considering the magnetic force in an air gap in the middle of a long bar-magnet, which is equivalent to the solenoid for this purpose. The force is $4\pi\sigma$, where $\sigma$ is the charge of magnetism per unit area of the face of the cut.

The strength $S$ of a shell is the magnetic moment per unit area, that is, $2ml/\alpha$ or $2l\sigma$. Hence $\sigma$ is $S/2l$, and, in the case of the current, is equivalent to (total current)$/2l$, that is, to the current per unit length of the solenoid or $cn_1$. Thus, as before, $H$ is $4\pi n_1 c$.

**40.** The expression $4\pi c$ for the work done in carrying a unit pole round a current enables us also to calculate the force due to a long, straight current. By considering the magnetic force near the edge of the equivalent shell, we see that the lines of force due to the current are circles surrounding the wire with their planes at right angles

*Magnetic force due to a long, straight current. Ampère's hypothesi*

to its length. These circular lines of force may be mapped out by a short compass-needle, or by sprinkling iron filings over a card threaded by the wire. By carrying a unit magnetic pole round the wire once, an amount of work equal to $4\pi c$ is done. The length of path of the pole is $2\pi r$, where $r$ is the radius of its orbit. The force is uniform over such a circular orbit by symmetry, and the magnetic force, which is

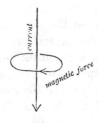

Fig. 54.

defined as the force on a unit north-seeking magnetic pole, at a distance $r$ from the wire, is given by

$$H = \frac{4\pi c}{2\pi r} = \frac{2c}{r}.$$

It was shown by Ampère that the magnetic effects of complete currents were the same as though each element of length $\delta l$ of the current produced its own magnetic force at a point equal to $c\delta l \sin \theta/r^2$, where $\theta$ is the angle made by the elementary length with the line joining it to the point at a distance $r$. The conception of an isolated length of current is not in accordance with Maxwell's view of the electric current, which he regarded as essentially a flow round a complete, closed circuit. Mr Heaviside, however, has evaded this difficulty by the conception of what he calls the "rational current element."

Let us imagine that a circuit is formed of a short straight element of current $\delta l$, and of lines of current-flow in the surrounding space; the lines being those mapped out by the lines of magnetic induction of a short bar-magnet. The current all passes along the element $\delta l$ in one direction, and returns by many paths spreading through the neighbourhood to form a closed circuit. This system constitutes Heaviside's rational current element. We must trace two properties of such elements:

(1) If two elements be placed together, the lines of current flow which formerly diverged from each end of one element will now pass along the second element before diverging, just as do the lines of induction if two short magnets be placed together to form a longer one. And, as a long magnet may be made up of any number of short ones, all poles disappearing except those at the ends, so current elements may be joined together, and all the lines of flow will pass through the elements in series before diverging from the ends of the chain. If we carry this process to its conclusion, and form a closed ring of current elements, all current will flow round the ring, and none will spread through the surrounding space—we have, in fact, the analogue of a complete ring of iron, when the lines of induction circulate round the ring and never leave the iron. We see, then, that a closed chain of rational current elements forms a closed circuit of the kind actually known in practice, when all the current flows within the circuit, and none spreads through the neighbouring medium.

(2) The second important property of the rational current element is the production of a magnetic force in accordance with Ampère's formula. Let $AB$ (Fig. 55) be the short length of the current element. Lines of return flow spread out from $A$ and

converge again to $B$. These lines are, by hypothesis, coincident
with the lines of induction given
by a short magnet lying at $AB$.

With its centre at $O$, a point
on the axis of $AB$ produced,
describe a circle, the plane of the
circle being normal to the axis
of $AB$. Let us calculate the total
magnetic induction through this
circle due to a magnet at $AB$.
The number of lines of induction
through $AB$ is equal to $c$ the
strength of current, and from
each end of the element $AB$ the

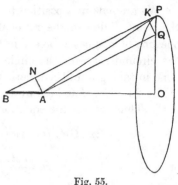

Fig. 55.

lines diverge uniformly. The number which threads the circle
round $O$ is measured by $c/4\pi$ per unit solid angle, $4\pi$ being the
measure of the solid angle which a complete sphere subtends at
its centre. If $\omega_A$ and $\omega_B$ denote the solid angles subtended by
the circle at $A$ and $B$ respectively, the total induction through
the circle is

$$\frac{c}{4\pi}(\omega_A - \omega_B),$$

and, by analogy, this denotes also the amount of current passing
through the circle when the hypothetical magnet is replaced by
the hypothetical rational current element.

Now join $A$ and $B$ to any point $P$ on the circumference of the
circle, and parallel to $BP$ draw $AQ$ to meet the radius $OP$ of the
circle in $Q$. Draw $QK$ perpendicular to $AP$. The quantity $\omega_A - \omega_B$
is the solid angle subtended at $A$ by the annular ring described
by the revolution of $PQ$ round the circle, and this solid angle is
sensibly equal to that subtended by the ring described by $QK$.
Hence

$$\omega_A - \omega_B = 2\pi \cdot OP \frac{QK}{r^2},$$

where $r$ denotes the distance $AQ$, which is sensibly equal to $AP$.
Draw $AN$ perpendicular to $BP$; we then have

$$\omega_A - \omega_B = 2\pi \cdot OP \frac{AN}{r^2}$$

$$= 2\pi \cdot OP \frac{\delta l \sin \theta}{r^2},$$

where $\theta$ is the angle between $AB$ and $BP$, and $\delta l$ is written for the element of length $AB$.

We are now in a position to calculate the magnetic force $H$ at the point $P$ due to the rational current element. By symmetry, one line of magnetic force due to the current must coincide with the circumference of the circle $OP$, and we know that the work done in taking a unit magnetic pole round that circle is measured (1) by $2\pi . OP . H$ and (2) by $4\pi$ times the current enclosed by the path. Thus, we get the equation

$$2\pi . OP . H = 4\pi \left( \frac{c}{4\pi} \, 2\pi OP \, \frac{\delta l \sin \theta}{r^2} \right),$$

or
$$H = \frac{c\delta l \sin \theta}{r^2},$$

in accordance with Ampère's formula, the magnetic force being at right angles both to $\delta l$ and $r$.

By the two properties (1) and (2) of Heaviside's rational current element, we have justified the use of Ampère's formula in calculating the magnetic forces due to complete current circuits of any size and shape. The resultant magnetic force of any circuit is equal to the sum of the forces due to its individual rational current elements.

We may therefore use Ampère's formula to calculate the magnetic effects of different currents, whether those currents be continuous ones carried in conductors, or the flights of charged particles which have become recently of so much importance, and will be considered in a later chapter.

Ampère's formula may also be applied to calculate the mechanical forces between circuits, or circuits and magnets. The magnetic force due to the current element being $c\delta l \sin \theta/r^2$, and that due to a magnetic pole of strength $m$ being $m/r^2$, it follows from the principle of equivalence between currents and magnets that $c\delta l \sin \theta$ may be written in place of $m$. Now if a pole $m$ be placed in a field where the magnetic force is $H$, the mecha-

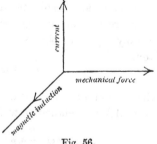

Fig. 56.

nical force on it is $Hm$. Thus on the current element the mechanical force is $Hc\delta l \sin \theta$ in air, or $Bc\delta l \sin \theta$ in any medium, the direction of the force being at right angles both to the current and to the lines of induction, and $\theta$ denoting the angle between the lines of induction and the direction of the current.

A long straight current produces at a distance $r$ a magnetic force $2c/r$. If a second long straight current $c'$ be brought near and parallel to the first, the mechanical force on a length $l$ must be $2cc'l/r$. Since this force acts at right angles both to the current and to the induction, it is a direct force of attraction between the currents if they flow in the same direction, and a direct repulsion if they flow in opposite directions.

**41.** For a circle of wire carrying a current $c$, Ampère's formula *A circular current.* for the magnetic force due to an element of current leads to a very simple result. The distance from the centre is everywhere uniformly $r$, and $\sin \theta$ is everywhere unity. For each element of current the magnetic force at the centre of the circle is therefore $c\delta l/r^2$, and, for the whole circumference,

$$\frac{2\pi cr}{r^2} = \frac{2\pi c}{r}.$$

This result gives a convenient means of putting our definition of unit current (p. 101) into a more practical form. Unit current is evidently that current which, when flowing in a circle of radius $r$, produces at the centre a magnetic force of $2\pi/r$.

The importance of this result makes it desirable that the magnetic force at the centre of a circular current should also be derived directly from the idea of the equivalence of currents and shells, without the use of Ampère's hypothesis.

Let $AB$ represent a circular current, and $O$ its centre. We will calculate the magnetic force at $P$, a point on the axis of the circle. The magnetic potential at $P$ is by § 38 equal to $c\Omega$, where $c$ is the current in electromagnetic units, equiva lent to the units of strength of magnetic shell, and $\Omega$ the solid angle subtended at the point by the circle.

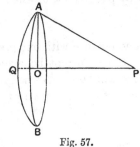

Fig. 57.

The solid angle is measured by the area of the spherical cap, drawn through $AB$ with $P$ as centre, divided by the square of the distance $AP$. The area of the cap is equal to that of a cylindrical ring of equal depth, cut from the cylinder which touches the sphere, of which the cap forms part, by two parallel planes, one containing the circle $AB$, and the other touching the top of the cap at $Q$, a point on $PO$ produced. Thus, the area of the cap is $2\pi \overline{AP} . \overline{OQ}$, and the solid angle which it subtends at $P$ is $2\pi \overline{AP} . \overline{OQ}/\overline{AP}^2$.

The magnetic potential at the point $P$ is given by the relation

$$V = 2\pi c \frac{\overline{AP} . \overline{OQ}}{\overline{AP}^2}.$$

Thus, if we denote the distance $OP$ by $x$, and $OA$, the radius of the current, by $r$, we have

$$V = 2\pi c \frac{\overline{AP}(\overline{AP} - x)}{AP^2}$$

$$= 2\pi c \left\{ 1 - \frac{x}{(r^2 + x^2)^{\frac{1}{2}}} \right\}.$$

Now the magnetic force $F$ in any direction $x$ is, by the definition of potential (§ 8), equal to the rate of decrease of the magnetic potential per unit distance in that direction, or $- dV/dx$. Thus,

$$F = - \frac{dV}{dx}$$

$$= 2\pi c \frac{r^2}{(r^2 + x^2)^{\frac{3}{2}}},$$

and this result gives us the magnetic force at any point on the axis of a circular current. It may also be derived directly from Ampère's formula. We can find the magnetic force at the centre $O$ by putting $x = 0$, when

$$F = \frac{2\pi c}{r},$$

in accordance with our previous result.

If the current be made to pass $n$ times round the circle, the force will be $n$ times as great; the magnetic force at the centre of a coil of $n$ turns of wire is given by

$$F = \frac{2\pi n c}{r}.$$

The calculation of the mechanical force between two circular currents requires mathematical analysis, except in simple cases.

One such case consists of two circles of equal radius $r$, placed parallel with a distance $d$ between their planes, $d$ being very small compared with $r$. We may then apply our result on page 105 for two long parallel currents, and obtain at once the expression $4\pi r c c'/d$ for the mechanical force between the two circles, $c$ and $c'$ being the respective currents.

**42.** We are now in a position to measure a current of electricity in absolute electromagnetic units, and it will be useful to consider exactly the meaning of such a measurement.

The tangent and ballistic galvanometers.

The proportionality between a current and the magnetic force it produces is a matter of definition. We do not *prove* that the magnetic force is proportional to the current, but we agree to *define* the strength of a current as the strength of the hypothetical magnetic shell, which may be imagined to replace the current, and to produce the same magnetic field. Then, in one of the two ways we have given, we calculate the magnetic force of the shell which is equivalent to a current flowing in a circular coil of wire.

We may now suspend a compass needle at the centre of such a coil, and place the plane of the coil in the magnetic meridian, so that the needle lies in that plane when no current passes. If the coil be large and the needle short, its ends will never be far from the centre of the coil, and the needle may be considered to lie in a uniform magnetic field, the direction of which is normal to the magnetic meridian and the plane of the coil. For convenience in observing, a light and long pointer is fixed to the needle at right angles to its length, or a mirror arrangement may be used. In accordance with the principles described

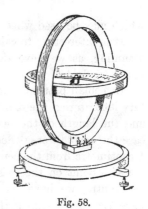

Fig. 58.

on p. 65, the needle will be deflected by this magnetic field $F$ through an angle $\theta$, which is given by the relation

$$F = H \tan \theta,$$

$H$ being the horizontal component of the earth's magnetic force.

Thus, if a current $c$ flow round $n$ turns of wire arranged in a large circular coil of radius $r$, we have

$$\frac{2\pi nc}{r} = H \tan \theta,$$

or

$$c = \frac{Hr \tan \theta}{2\pi n}.$$

For accurate work, it is necessary to take account of the fact that all the turns of wire do not lie in a single circle. For rough work, $r$ may be put equal to the mean radius of the coil.

The quantity $2\pi n/r$ depends only on the dimensions of the galvanometer, and, if we call this quantity the galvanometer-constant and denote it by $G$, we get

$$c = \frac{1}{G} H \tan \theta.$$

If we keep the instrument always fixed in one position in the laboratory, and no masses of iron are moved in the neighbourhood, the local value of $H$ may be taken as approximately constant and can be found once for all. We may write all the constants of our equation as a single *reduction factor k*, and obtain the simple formula

$$c = k \tan \theta,$$

which may be used when no high degree of accuracy is needed.

It is interesting to calculate the percentage error involved in assuming, as we have done, that the needle lies in a uniform magnetic field.

In a small pattern tangent galvanometer, often seen in elementary laboratories, the diameter of the coil is about 20 centimetres, and the length of the needle about 2 centimetres. If in such a case the needle were deflected through 90°, its ends would lie each 1 centimetre from the centre of the coil.

At a point on the axis of a circular coil, at a distance $x$ from the centre, we have seen (p. 110) that the magnetic force is $2\pi cr^2/(r^2 + x^2)^{\frac{3}{2}}$, while at the centre it is $2\pi c/r$.

Now $r^2/(r^2 + x^2)^{\frac{3}{2}}$, in this case, is equal to $100/(101)^{\frac{3}{2}}$ or about $100/1015$, while $1/r$ is $1/10$. Thus the percentage difference is about 1·5, and we see that, with the rough instrument we have described, results accurate to about one and a half per cent. should be obtained if the coils may be assumed to lie in a single circle.

The accuracy can of course be increased by using larger coils or shorter needles. Still more exact results can be obtained by placing two or three coils parallel to each other so that the coils lie on the surface of a sphere at the centre of which the needle is suspended.

The tangent galvanometer gives us our first practical means of measuring a current in absolute, electromagnetic units. As a practical unit, a current equal to one-tenth of the electromagnetic unit is chosen, and called the ampere.

It is sometimes necessary to measure, not the strength of a steady current, but the whole quantity of electricity which passes during the time of flow of a transient current, such as is obtained by the discharge of a condenser, or by electromagnetic induction. A galvanometer adapted for this purpose must have a small needle, so that it lies in a uniform field of force. The moment of inertia of the suspended system must be high, and the time of swing great, so that all the current has passed before the needle has moved appreciably from its position of equilibrium. An impulse is thus given to the needle, and, by the usual mirror arrangement, the extreme limit of its first swing or throw may be observed. The instrument is called a ballistic galvanometer, and the following investigation gives the theory of its action.

Let $ACB$ be the position of equilibrium of the magnet, and $A'CB'$ the limit of its swing through an angle $\beta$ (Fig. 59). Draw $A'D$, $B'E$ at right angles to $ACB$. Then the lengths $AD$ and $BE$ represent the equal distances through which the poles of the needle are moved against the direction of the horizontal magnetic force $H$ of the earth. If $m$ and $-m$ are the pole strengths, the work done is $2mH \cdot AD$. But if $l$ be the half length of the needle,

$$AD = CA - CD = l(1 - \cos \beta),$$

and the work done by the earth's field in stopping the swing is

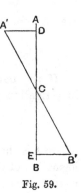

Fig. 59.

$$2mHl(1 - \cos \beta) = MH(1 - \cos \beta),$$

where $M$ is the magnetic moment of the needle. This work must be equal to the kinetic energy of the needle at starting on its swing, *i.e.* to $\frac{1}{2}K\omega^2$, where $K$ is the moment of inertia, and $\omega$ the initial angular velocity. Hence

$$\omega = \sqrt{\left\{ \frac{2MH(1-\cos\beta)}{K} \right\}}.$$

Now the couple produced on the needle by the action of the current $c$ is $MGc$, where $G$ is the galvanometer constant, and, if the current last for a time $\delta t$, the impulse is $MGc\,\delta t$. For the whole time of flow of the transient current, the total impulse is then $MG\int c\,dt$. Now $c\,\delta t$ is the amount of electricity passing during the time $\delta t$, and $\int c\,dt$ is the total amount $q$ of electricity passing through the galvanometer. Thus the impulse on the needle is $MGq$, and this must be equal to the moment of momentum $K\omega$ of the needle. Hence $\omega$ is equal to $MGq/K$.

We now have

$$\frac{MGq}{K} = \sqrt{\left\{ \frac{2MH(1-\cos\beta)}{K} \right\}},$$

and thus

$$q = \frac{2\sin\frac{1}{2}\beta}{G}\sqrt{\left(\frac{HK}{M}\right)}.$$

If there be no appreciable friction on the needle, and consequently no appreciable damping of the swing, the period $T$ of one complete oscillation is

$$T = 2\pi\sqrt{\frac{K}{MH}}.$$

Hence, substituting for $K/M$ in the equation for $q$ we get

$$q = \frac{HT}{G\pi}\sin\frac{1}{2}\beta.$$

If the damping be appreciable, the complete mathematical investigation shows that

$$q = \frac{HT}{G\pi}\left(1 + \frac{\lambda}{2}\right)\sin\frac{1}{2}\beta,$$

where $\lambda$ is the so-called logarithmic decrement, that is, the natural logarithm of the ratio of the amplitude of one swing to that of the next.

A suspended coil galvanometer can also be used ballistically. The force on each vertical side of the coil is $cHl$, where $l$ is the length of each side, $H$ the strength of magnetic field in which the coil hangs, and $c$ the total current passing through the side, that is the current multiplied by the number of turns of wire.

The total impulse is $Hlq$, and the moment of momentum of the coil about the axis of suspension is $Hll'q$, where $l'$ is the length of a horizontal side of the coil. But $ll'$ is the area $A$ of the coil. Hence
$$HAq = K\omega.$$

The coil is brought to rest at a deflection $\beta$ by the work it has to do in twisting the suspending wire or strip. The couple produced by the twist is proportional to the angle of deflection, hence, if $b$ be the couple for unit twist, the work done for an additional small twist $d\beta$ is $b\beta d\beta$, and for the whole deflection is $\int_0^\beta b\beta d\beta$ or $\frac{1}{2}b\beta^2$. Therefore the initial kinetic energy $\frac{1}{2}K\omega^2$ is equal to $\frac{1}{2}b\beta^2$, and $\omega^2$ is $b\beta^2/K$. But we also have that $\omega^2$ is $H^2A^2q^2/K^2$. Thus we get
$$q^2 = \frac{b}{H^2A^2} \cdot K\beta^2.$$

The time $T$ of one oscillation of the coil is $2\pi\sqrt{\dfrac{K}{b}}$. Hence
$$q = \frac{bT}{2\pi HA}\beta.$$

With either type of galvanometer, the constants of the instrument can be determined by passing a steady current of known strength through it.

**43.** In dealing with the phenomena of electrostatics, we had occasion to introduce the conception of electric potential, and we defined the difference of potential between two points as the work done against the electric forces when unit quantity of electricity was carried from one point to the other.

Electromotive force.

When an electric current flows along a wire, we conceive that electricity is passing, and thus the two points along the wire must be maintained by some means at a permanent difference of potential. We may, if we please, consider the maintenance of this potential-difference as the function of the electric battery or other source of current.

In the science of current-electricity, it is usual to call a difference of potential an *electromotive force*. The name is not a

happy one, for an electric force $f$ has the physical dimensions [force/quantity of electricity], and is related to potential-difference $V$, or [work/quantity of electricity], by the equation
$$f = -dV/dx.$$
The use of the name electromotive force as a synonym for difference of potential, however, is established firmly, and the convention will be accepted in this book.

The electromotive force between two points in a circuit, then, is defined as the work done by the electric forces when one electromagnetic unit of electricity passes from the point at higher to the point at lower potential.

To connect the electromotive force with the current, another definition is needed. On the conception of a current by which we regard it as analogous to the flow of an incompressible fluid, in § 35, we have already defined the electromagnetic unit of electricity as that quantity which we must suppose to pass each cross-section of a circuit when unit current flows round the circuit for unit time. Thus it follows that our definition of the electromotive force between two points is equivalent to saying that the electromotive force is represented by the work done by the electric forces when unit current passes between the two points for unit time.

On the system of units based on the centimetre, gramme and second (C.G.S. system) the unit of work is the erg. The unit electromotive force, then, is the difference of potential which exists between two points if one erg of work is done when one electromagnetic unit of current flows between the points for one second. We shall consider the experimental determination of this unit in the sequel. It will be found to be inconveniently small for ordinary purposes, and a practical unit called a *volt*, which is $10^8$ absolute or electromagnetic units, is employed.

Differences of potential, or electromotive forces, can be compared by electrostatic means. If the potential differences are great, they may be demonstrated by a gold-leaf electroscope, and even the small electromotive force of a voltaic cell may be measured fairly accurately in arbitrary units by means of a quadrant electrometer.

The electromotive force of a cell as thus measured will be found to depend on the materials and nature of the cell only, not on the dimensions of the plates or the amount of liquid used. If several cells be joined together in series, with the zinc plate of one cell clamped to the copper of the next and so on, the plates clamped together must be at the same potential. Thus, if the negative terminal of the first cell be joined with the earth, and be taken as at zero potential, the negative pole of the second cell must be at the potential of the positive pole of the first, that is, at a potential $e$, where $e$ is the electromotive force of each cell. The difference of potential between the plates of each cell being the same, in passing from the negative to the positive pole of the second cell the potential again arises by $e$, and the potential of the positive pole of the second cell is $2e$. Similarly, if $n$ cells be joined together in series, the potential of the positive terminal of the $n$th cell is $ne$, and this is the value of $E$ the effective electromotive force of the battery of cells. The conclusion may be confirmed by experiments with a quadrant electrometer. We now see that, by increasing the number of cells in series, any required electromotive force may be applied to a given circuit.

**44.** The first to introduce exact ideas on this subject was Dr G. S. Ohm, who in 1827 replaced the prevalent vague notions of "quantity" and "intensity" by the definite conceptions of current-strength and electromotive force. He also stated the law called by his name, which expresses the experimental result that the current-strength between two points of a circuit is directly proportional to the applied electromotive force. Ohm verified his ideas by experiments with voltaic cells and thermoelectric piles (see § 52), and found that, along a homogeneous linear conductor, the rate of fall of potential is constant.

Ohm's law, that the current $c$ varies as the applied electromotive force $E$, may be stated in the form

$$c = kE,$$

where $k$ is a constant known as the conductivity of the conductor. Another mode of statement, more usually adopted, is to say that

$$c = \frac{1}{R} E,$$

or

$$E = Rc,$$

where $R$ is a constant which, like the conductivity, depends only on the nature, dimensions and temperature of the conductor, and is independent of the current. This constant $R$ is called the resistance.

Lightning has been known to melt conductors which it strikes, and these effects were imitated on the small scale by the sparks from electric machines by Franklin and others. At an early date in the history of current electricity, it was noted that heat was developed in wires connected with the two poles of voltaic batteries, and in 1821 Sir Humphry Davy described to the Royal Society experiments which showed that the amount of heat liberated varied greatly with the nature of the metal. The development of heat was referred to the resistance offered by the wire to the passage of the current, and the relative amounts of heat produced by the same current in different conductors were taken as inversely proportional to their relative conducting powers. In 1826 similar experiments were made by Sir W. Snow Harris with discharges of statical electricity obtained from a battery of Leyden Jars.

The definite conceptions of current and electromotive force introduced by Ohm led to an equally definite idea of electric resistance, as a constant depending only on the nature, dimensions and temperature of the conductor. It became possible to compare two resistances by applying to them the same electromotive force, and measuring the relative strengths of the resulting currents.

The unit of resistance is evidently to be defined as the resistance through which unit electromotive force will maintain unit current. If electromagnetic units of current and electromotive force be used, the electromagnetic unit of resistance follows. With practical units, through unit resistance, one volt will maintain a current of one ampere. This resistance is called the ohm, and is $10^9$ absolute electromagnetic units.

With the limit of accuracy at present possible, the ohm has been found to be equivalent to the resistance of a column of mercury of 1 square millimetre in cross section and 106·3 centimetres in length, the temperature being that of melting ice.

This conducting system is easily reproduced, and, for convenience, a practical definition of the international ohm has been adopted on these lines. To avoid certain difficulties of measurement, the definition runs thus :—A column of mercury of uniform cross section, 106·3 centimetres in length, and of mass 14·4521 grammes, at the temperature of melting ice, has a resistance of one ohm.

Even should increased accuracy of experiment show differences between the true theoretical ohm and this definition, it is probable that the definition will still be kept as describing the practical unit.

Standard coils of one or more ohms are now made, and their error determined by the National Physical Laboratory. Figure 60 shows one form of standard ohm coil. The wire is bare, and made of a platinum-silver alloy. It is wound on a mica frame, and immersed in an insulating oil, of which the temperature is observed with a thermometer. This arrangement insures that the coil should be at the indicated temperature, and, in this respect, is much better than the

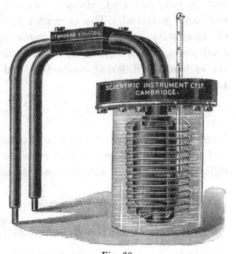

Fig. 60.

more common method of coating resistance coils with silk, paraffin or other solid insulators, which are bad thermal conductors. The leads connected with the ends of the coil are made of thick copper bars, which may be joined to an electric circuit by means of mercury cups or, still better, cups filled with fusible alloy.

In an improved form, due to Dr Rosa, the coil is immersed in dry oil hermetically sealed in a metal case. The coil has separate terminals for current and potential, the former being binding-screws of very large contact surface.

The conception of electric current as a continuous flow of electricity through the substance of the conductor, combined with the experimental relation known as Ohm's law, enables us to predict the resistances of systems of two or more conductors arranged in different ways.

If a series of conductors, $AB, CD, EF, \ldots$ (Fig. 61) be connected together in series, the end $B$ of $AB$ will be at the same potential as

Fig. 61.

the end $C$ of $CD$ clamped to it, and similarly with the ends $D$ and $E$, etc., provided, always, that the conductors are made of the same material. The current $c$ must, since no accumulation of electricity occurs, be the same throughout the series of conductors, and, if $E_1, E_2, \ldots$ be the potential-differences between the ends of the different conductors, and $r_1, r_2, \ldots$ their respective resistances,

$$ c = \frac{E_1}{r_1} = \frac{E_2}{r_2} = \ldots $$

Considering the whole series as one conductor, with a resistance $R$ and a potential difference of $E$, equal to the sum of the potential differences $E_1, E_2, \ldots$ etc., between its ends, we have also

$$ c = \frac{E}{R}. $$

Now $\qquad E = E_1 + E_2 + E_3 + \ldots,$

Thus $\qquad cR = cr_1 + cr_2 + cr_3 + \ldots,$

or $\qquad R = r_1 + r_2 + r_3 + \ldots.$

Another important case occurs when the conductors have all their beginnings connected together and all their ends connected together, as in Fig. 62. With this arrangement, the conductors are said to be in parallel or multiple arc. The current $c$ flowing in

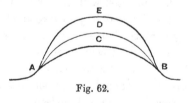

Fig. 62.

at $A$ must, by our principle, be the sum of the currents $c_1, c_2, \ldots$ in the branches $C, D, \ldots$. The difference of potential $E$ between the ends of each branch is obviously the same. We have

$$c = c_1 + c_2 + c_3 + \ldots$$

$$= E \left( \frac{1}{r_1} + \frac{1}{r_2} + \frac{1}{r_3} + \ldots \right).$$

Considering the whole system between $A$ and $B$, we have

$$c = \frac{E}{R}.$$

Thus
$$\frac{1}{R} = \frac{1}{r_1} + \frac{1}{r_2} + \frac{1}{r_3} + \ldots.$$

The reciprocal of the resistance of a conductor is called its conductivity, and we may express our last result by saying that the conductivity of a number of conductors arranged in parallel is the sum of their individual conductivities.

These results for conductors in series and parallel may be verified experimentally, and thus a confirmation be obtained of Ohm's law and of the justness of the conception of the current as analogous to the flow of an incompressible fluid through the conductors.

By measuring the resistances of wires of different diameters but of the same material, it is found that, with direct continuous currents, the resistance of a wire varies inversely as its area of cross section. This follows also theoretically from the principle of the summation of conductivities if we imagine a number of the wires first arranged in parallel and then to coalesce. It follows that the current flows through the substance of the conductor and not over its surface. This result, however, does not hold for rapidly alternating currents.

The principles of continuous current-flow which we have now established may conveniently be applied to complex circuits and networks of conductors in the form of two statements known as Kirchhoff's laws.

1. The algebraical sum of the currents which meet at any point is zero.

2. In any closed circuit the algebraical sum of the products of the current and resistance in each of the conductors in the circuit is equal to the electromotive force in the circuit.

The first of these laws expresses the result that there is no accumulation of electricity anywhere in the circuit, the second follows from Ohm's law as applied to each complete circuit to be found throughout the network.

**45.** A very exact method of applying Ohm's law to the Wheatstone's comparison of resistances was introduced by Christie bridge. and Wheatstone about 1843. The apparatus is known as Wheatstone's bridge.

The circuit of a voltaic cell (Fig. 63) is made to branch at $A$ into two arms $ACB$ and $ADB$, which rejoin each other at $B$. Let the current flow from $A$ to $B$, so that the potential $V_A$ at the point $A$ is higher than the potential $V_B$ at the point $B$.

As we pass along

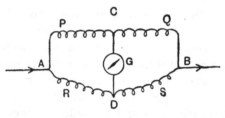

Fig. 63.

the arm $ACB$, the potential falls from $V_A$ to $V_B$, and the potential $V_C$ at $C$ must have some value intermediate between $V_A$ and $V_B$. But, along the lower arm $ADB$ the same fall of potential occurs, so that there must be some point $D$ in the lower arm which has the same potential as the point $C$ in the upper arm.

If one terminal of a galvanometer be connected with $C$, and the wire from its other terminal be moved along $ADB$ till it comes to $D$, the terminals will then be at the same potential, and no current will flow through the galvanometer. When this is the case, we know that $V_C$ is equal to $V_D$.

When no current flows along the cross connexion $CD$, the current $c_1$ in the arm $AC$ must all pass out along the arm $CB$, and the currents in these arms be the same. The current $c_2$ in the arm $AD$ must be equal to the current in the arm $DB$.

Let $P, Q, R,$ and $S$ denote the resistances of the four conductors $AC, CB, AD,$ and $DB$ respectively. Then, by Ohm's law, we know that, for each conductor, the electromotive force is equal to the product of the current and the resistance, and we get the relations

$$V_A - V_C = Pc_1, \text{ and } V_C - V_B = Qc_1,$$

or
$$\frac{V_A - V_C}{V_C - V_B} = \frac{P}{Q}.$$

Similarly
$$\frac{V_A - V_D}{V_D - V_B} = \frac{R}{S},$$

and, since $V_C = V_D$ when no current passes through the galvanometer, we have

$$\frac{P}{Q} = \frac{R}{S}.$$

Thus, if we know $S$ and the ratio $P/Q$, we can calculate the value of the resistance $R$. If $S$ be constructed to be a definite number of ohms, $R$ is known in ohms also.

The practical application of these principles is carried out in many ways. In the wire bridge, a long thin uniform wire, preferably of platinum-silver or some other alloy not easily oxidised, is stretched alongside a scale, as shown in Figure 64.

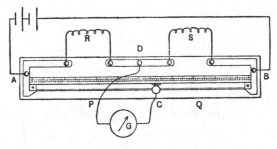

Fig. 64.

The ends of this wire are soldered to thick copper bars of negligible resistance, and a coil of known resistance is inserted in the gap in the bar at $R$. The coil to be compared with that in $R$ is inserted in the gap $S$. A galvanometer is connected with the screw $D$, between $R$ and $S$, and with a travelling jockey, which, by the pressure of a key, makes sharp contact with the wire at $C$. The position of the jockey is varied till, on pressing the key, no current flows through the galvanometer. When this is the case, $S$ is given by the quantity $RQ/P$, where $R$ and $S$ again denote the resistances of the corresponding coils. It is easy to show that the sensitiveness of this arrangement is greatest when $R$ and $S$ are nearly equal, and the jockey consequently near the middle of the bridge wire. Thus, a coil should be selected for insertion in $R$ of approximately the same resistance as the coil to be compared with it.

For a second type of measurement, an arrangement of coils known as a resistance box is required. The brass blocks $A$ and $B$ in Figure 65 are connected inside the box through a coil of wire of known resistance. The coil is doubled on itself and wound on an insulating

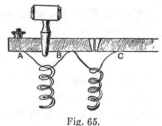

Fig. 65.

bobbin in this way in order to avoid magnetic effects on the galvanometer and the effects of self-induction (§ 55). The resistance of the coil may be cut out of the circuit by inserting a brass plug as shown between $A$ and $B$. With a series of such coils any required resistance can be thrown into or taken out of the circuit.

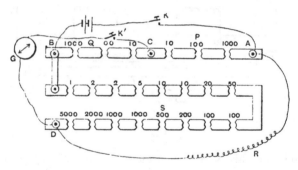

Fig. 66.

One arrangement of coils, sometimes known as a Post-office box, is shown in Fig. 66. Plugs are taken out of $P$ and $Q$ so as to give a convenient ratio, and $S$ is then adjusted till the bridge is balanced, when $R$ can be calculated. By making the ratio of $Q$ to $P$ equal to 10 to 1, a single ohm coil in $S$ is equivalent to 0·1 of an ohm in $R$. Thus fractions of an ohm may be measured.

A second and, in some ways, more convenient arrangement of coils is the box shown in Figure 66 $a$. The ratio arms are similar to those of the Post-office box, but the arm $S$ is made up of coils placed in groups of units, tens, hundreds, and thousands.

In the unit set, single ohm coils join the sectors marked 1, 2, 3, etc. to each other in series. By inserting the plug opposite 3, for instance, three ohm coils are put between the central brass bar of

Fig. 66 a.

the set, and the zero brass, which forms the other terminal. Any resistance up to 9999 ohms can be put into the arm $S$ by changing the position of the four plugs in the dials.

In more modern types of resistance box the contacts are made by switches which travel round a dial. Between each segment of

Fig. 67.

the dial is fixed a coil of equal resistance. Thus as the switch travels round, it puts into circuit resistances rising from 1 to 10. Usually four dials are supplied, with units, tens, hundreds and thousands of ohms. The ratio coils are inserted by plugs. To get good contact the switch carries brushes which are made of hard copper with a large number (*e.g.* 24) of laminations each of which makes contact both with a central flange and with the dial.

A modification of the wire-bridge arrangement, due to Professor Carey Foster, enables us to find the difference in resistance between two coils with extreme accuracy. A stretched wire (Fig. 68) is soldered to copper bars as in the usual form of bridge, but, instead of two gaps in the bars, there are four. In the two middle ones, on each side of the galvanometer connexions, are placed two

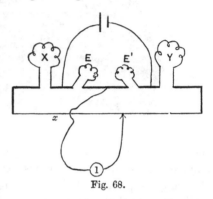

Fig. 68.

nearly equal coils $E$, $E'$, kept in the same conditions and at the same temperature. In the two outer gaps, beyond the battery connexions, are placed the two coils the difference of which is required; let us call their resistances $X$ and $Y$. Let us adjust the jockey along the wire till the bridge is balanced. Let the length of wire to the left of the jockey then be such that its resistance is $x$. Now suppose that we interchange the two coils $X$ and $Y$, and find a new position of balance for the jockey, such that the resistance of the wire to the left of the jockey is $y$. The two nearly equal arms, which form one pair of arms for the bridge, being unchanged, the resistances of the other two arms of the bridge must be unchanged also, and thus the new arm to the left, namely $Y + y$, must be identical in resistance with the old arm in the same position, namely $X + x$. Thus

$$X - Y = y - x,$$

or the difference of resistance between the two coils is equal to the resistance of the length of wire between the two positions of equilibrium of the jockey. This resistance can be found with great accuracy by measuring the total resistance of the whole length of the wire, and calibrating it throughout.

**46.** The resistance of a conductor as hitherto considered
Specific depends on the shape and dimensions of the con-
resistance. ductor as well as on its nature and temperature. It
is evidently convenient to form the conception of some quantity
which depends only on the nature and temperature of the substance
conducting, and not on the particular shape or dimensions of the
particular specimen employed. It is usual to take the resistance
between opposite faces of a cube of the substance, with a length of
side of one centimetre, as the specific resistance of the material.
It would, of course, usually be impossible, and always inconvenient,
to measure the resistance of such a body directly. Its resistance
can, however, be calculated from that of wires, or other convenient
masses of the substance, by the aid of the results given on page 121,
which show that the resistance of a wire or other conductor of
regular form is directly proportional to its length, and inversely
proportional to its area of cross section.

The following table of specific resistance is taken chiefly from
results given by Fleming and Dewar in 1893. The specific resistances are expressed in microhms per centimetre cube, a microhm
being the millionth, or $10^{-6}$, of an ohm. A second column gives the
resistance in ohms of a column of the material 1 metre in length
and 1 square millimetre in cross section.

|  | Specific resistance in microhms per centimetre cube | Resistance in ohms of column 1 metre by 1 sq. mm. | Mean temperature coefficient per degree between 0° and 100° C. |
|---|---|---|---|
| Silver, annealed ... | 1·468 | ·01468 | ·00400 |
| „ hard drawn ... | 1·615 | ·01615 | ... |
| Gold, annealed ... | 2·036 | ·02036 | ·00377 |
| Zinc ... ... ... | 5·751 | ·05751 | ·00406 |
| Copper, annealed ... | 1·562 | ·01562 | ·00428 |
| „ hard drawn | 1·603 | ·01603 | ... |
| Iron ... ... ... | 9·065 | ·09065 | ·00625 |
| Platinum ... ... | 10·917 | ·10917 | ·00367 |
| Mercury ... ... | 94·073 | ·94073 | ·00072* |

* This figure gives the approximate temperature coefficient of mercury near
20° C., as determined by Matthiessen.

The resistance of alloys is, in general, much higher than that
of pure metals, the effect of a very small admixture being sometimes surprisingly great. The change with temperature in the
resistance of alloys is usually less than that of pure metals; this

small change is an advantage in the construction of resistance-coils, etc., though it is seldom worth while to obtain a low temperature coefficient at the cost of permanence in properties, a quality in which some alloys seem to be deficient. Fleming and Dewar give the following figures:

| Alloys | Compositions in per cents. | Specific resistance | Temperature coefficient at 15° C. |
|---|---|---|---|
| Platinum-silver ... | Pt 33, Ag 66 ... | ... 31·582 | ·000243 |
| Platinum-iridium .. | Pt 80, Ir 20 ... | ... 30·896 | ·000822 |
| Platinum-rhodium ... | Pt 90, Rd 10 .. | .. 21·142 | ·00143 |
| German-silver ... | Cu 50, Zn 31, Ni 19 | ... 29·982 | ·000273 |
| Platinoid ... ... | German-silver with a little tungsten ... | ... 41·731 | ·00031 |
| Manganin ... ... | Cu 84, Mn 12, Ni 4 | ... 46·678 | ·0000 |

It should be stated here that the resistance of all metallic bodies depends not only on their chemical composition, but also on their physical state. Differences in processes of annealing, etc., which affect the crystalline structure of the metal, are found also to change the specific resistance. Tables of specific resistances and temperature coefficients, then, should only be regarded as strictly applicable to the particular specimens used by the experimenters, though, as approximate guides, such tables may be of general use.

The resistance of metallic conductors, unlike that of electrolytes to be studied hereafter, increases with rising temperature. The most recent and complete experiments on the resistance of metals throughout a wide range are those of Dewar and Fleming, who carried their observations to very low temperatures by the use of liquid air and other liquefied gases. The diminution of resistance at these low temperatures is very marked, and, when curves were drawn, it seemed as though the resistance of all pure metals would vanish in the neighbourhood of the absolute zero. Further experiments, however, at the extreme cold of liquid hydrogen, indicate that, before such temperatures are reached, the curves tend to become parallel to the axis—the resistances to approach a limiting value.

The resistance of wires of pure platinum, and its variation with temperature, have been the subject of many experiments. It was shown by Callendar that, if such coils of wire be placed in porcelain tubes or protecting cases made of glass, and carefully preserved from all strain by loose winding, the resistance was a function of the temperature only. At a given temperature, the resistance always had a definite value, irrespective of the treatment to which the wire had been subjected previously. This property of consistency is the first and last essential for a trustworthy thermometer, and platinum resistance thermometers are now used extensively not only for high and low temperatures, where other methods are impossible, but also at ordinary temperatures, in cases when the difference of two temperatures is required with extreme accuracy.

Figure 69 shows a platinum thermometer, and Figure 70 shows the form of wire-bridge which is now usually employed to

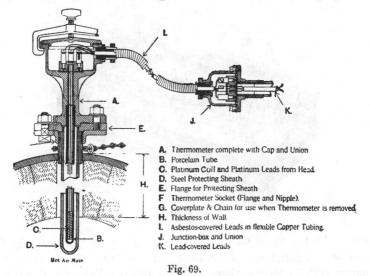

A. Thermometer complete with Cap and Union
B. Porcelain Tube.
C. Platinum Coil and Platinum Leads from Head.
D. Steel Protecting Sheath
E. Flange for Protecting Sheath
F  Thermometer Socket (Flange and Nipple).
G. Coverplate & Chain for use when Thermometer is removed.
H. Thickness of Wall.
I. Asbestos-covered Leads in flexible Copper Tubing.
J. Junction-box and Union
K. Lead-covered Leads

Fig. 69.

measure its resistance. The electrical arrangements are a modifi-cation of those of Carey Foster's bridge, and will be understood by a reference to Figure 68 on page 126. $E$ and $E'$ are two equal arms as before. The platinum thermometer is placed in the position marked $Y$, and in the position marked $X$ is inserted a coil of wire made of some substance with a negligible temperature coeffi-cient, or else a coil kept at a constant temperature. In order to allow for the variation with temperature of the conducting leads in the platinum thermometer—a variation impossible to calculate —a pair of dummy leads are fixed alongside the real leads in the tube. These compensating leads are connected together below by a short piece of platinum wire. Whatever irregular heating or cooling happens to the real leads, happens also to the dummy leads, which consequently suffer a change of resistance equal to that of the real leads. The dummy leads are inserted in the arm $X$ of the bridge, and thus compensate automatically the variations of resistance of the real leads in the arm $Y$. As the temperature of the thermometer is raised, the position of equilibrium of the bridge moves along the stretched wire from left to right. The range of the instrument can be increased by adding to the left part of the bridge-wire the resistance coils shown in Figure 68.

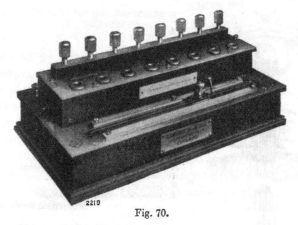

Fig. 70.

From the theory of Carey Foster's bridge, it follows that equal increases of resistance of the platinum coil correspond with equal movements in length along the bridge-wire. Thus, by dividing the wire into equal lengths, we obtain a scale of temperature degrees, which we may call the platinum scale, in which temperature is defined as proportional to the resistance of a platinum wire. The possibility of thus dividing the wire into equal lengths, corresponding to equal increments of temperature, constitutes the advantage of this form of bridge for the present purpose. If the platinum thermometer were inserted in the arm $S$ of the original form of Wheatstone's bridge, shown in Figure 64 on page 123, no such possibility would exist.

Callendar has shown that the platinum temperatures as thus obtained may be reduced to the scale of the gas thermometer by

means of the following empirical parabolic formula for the difference between the two scales:

$$t - t_p = \frac{t\delta\,(t - 100)}{10000} = \delta\left\{\left(\frac{t}{100}\right)^2 - \frac{t}{100}\right\},$$

where $t$ is the temperature on the gas scale, $t_p$ the temperature on
the platinum scale, both expressed in centigrade degrees, and $\delta$ is
a "difference-constant" the value of which for pure platinum is
about 1·5, but varies slightly for different specimens. It can be
determined by standardizing an instrument in ice, steam, and the
vapour of sulphur, which, at normal atmospheric pressure, boils at
a temperature of 444·5° C. on the scale of the gas thermometer.

The platinum thermometer is an admirable instrument for
research-work in physical laboratories. For technical purposes, its
advantages comprise the possibility of measuring the temperature
from a distance, the extended range of the scale, and the permanent
record of varying temperatures which it is now possible to obtain,
but it is now less used than thermo-electric thermometers described below. For scientific research the resistance thermometer
has been used both at high and low temperatures: for determining
the melting points of metals and the boiling points of liquefied
gases. The extreme sensitiveness of modern null methods of
measuring resistance renders the platinum thermometer a most
delicate instrument for estimating the differences between two
neighbouring temperatures—a difference of the ten-thousandth
part of a centigrade degree may be detected with some ease, while
careful experiments by Dr E. H. Griffiths have shown that the
hundred-thousandth part of a degree may be estimated.

**47.** As already stated, electromotive forces may be compared
electrostatically, roughly by a gold-leaf electroscope,
more accurately by a quadrant electrometer. But,
in practice, when the electromotive forces do not
exceed a few volts, it is usually more convenient to employ some
means based on the application of Ohm's law.

Comparison
of electro-
motive forces.

The currents produced successively in the same high resistance
circuit by two electromotive forces may be estimated by the
deflections of a galvanometer or of a voltmeter, which is usually
a moving-coil galvanometer of high resistance. By a second arrange-

9—2

ment, the resistance of two circuits may be varied till the same
deflection is produced in each case. In these methods, a current is
allowed to flow through the cell or other source of electromotive
force, and, to obtain the total electromotive force acting round the
circuit, the resistance of the cell must be added to that of the rest
of the circuit.

A method, which does not involve a knowledge of the resistance
of the cell, is based on
the use of an instrument
known as the potentio-
meter. A long, thin wire,
of high resistance, is
stretched by the side of
a scale (Figure 71).
Through this wire passes
a constant current, main-

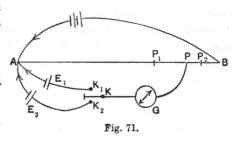

Fig. 71.

tained by a battery of cells of constant electromotive force greater
than that of the cells to be compared. The potential rises uniformly
as we pass along the wire from the end $A$ to the end $B$. Thus, if
one of the cells to be compared be connected with $A$ as shown,
and its other terminal be connected through a galvanometer with
different points along the wire, some point can be found at which
the potential is the same as that of the applied pole of the cell.
When this is the case, no current flows through the galvanometer,
and the potential-difference between $A$ and the point $P$ is equal
to the electromotive force of the cell. If the second cell be now
inserted in place of the first, a new position of equilibrium will
be found, the travelling wire being applied at the point $P_1$. The
electromotive force of this second cell must then be equal to the
potential-difference between the points $A$ and $P_1$. But, by applying
Ohm's law to the main circuit $AB$, we see that the differences of
potential between $AP$ and $AP_1$ respectively are proportional to
the resistances of the two lengths of wire $AP$ and $AP_1$. If the
wire be uniform, these resistances are proportional to the lengths,
and thus the ratio between the electromotive forces of the two
cells is equal to the ratio of the two lengths $AP$ and $AP_1$.

It is, comparatively speaking, easy to construct a standard
resistance-coil having a resistance equal to the practical unit or

ohm; but we cannot construct at will a cell having an electromotive force of one volt. We can only take the most constant cell we know, measure once for all its electromotive force in volts, and use the cell thereafter as an arbitrary standard of electromotive force.

One such standard cell is that invented by Latimer-Clark and shown in Figure 72. It consists essentially of mercury in contact with a paste of mercurous sulphate, on which rests a solution of zinc sulphate kept saturated by the presence of crystals of the salt. Into this solution dips an amalgamated rod of pure zinc. This cell must not be used to yield a current; but its electromotive force is very constant when no current is taken, and has the value of 1·432 or 1·433 volts at 15° C. The electromotive force diminishes by 0·00077 volt for each centigrade degree the temperature rises above 15°.

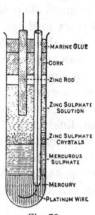

Fig. 72.

The Weston cell consists of the arrangement ·

mercury/mercurous sulphate/cadmium sulphate solution/cadmium.

Its electromotive force is 1·018 volt, at 15° C., and its temperature coefficient is considerably less than that of the Clark cell.

The principle of the potentiometer is applicable to a number of different purposes. It may, for instance, be used to measure a large current in the following manner. The current is passed through a strip of metal of small resistance (Fig. 73). Between

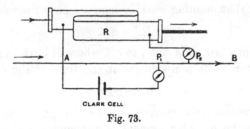

Fig. 73.

two potential terminals on the strip a difference of potential $cR$ must then exist, and this potential-difference may be compared with the electromotive force of a standard cell, along the wire of the potentiometer $AB$. Thus by previously standardising the apparatus $c$ may be determined.

A simpler method, more convenient in practice, consists in connecting the terminals of a voltmeter to the potential terminals of the metal strip $R$. The resistance of the voltmeter is so adjusted that its readings give directly the number of amperes of current flowing through the strip of metal. The voltmeter is thus converted into an ampere-meter or ammeter.

A similar process enables us to compare two resistances which are too small to be examined accurately by the usual means. The same current is passed through the two resistances in series, and the differences of potential between their ends compared by the potentiometer or a voltmeter consisting of a galvanometer of very high resistance or an electrostatic voltmeter such as those described on pp. 48, 49. These potential-differences will be proportional to the two resistances.

**48.** By the definition of electromotive force, it follows that the

**Heating effect of an electric current.**

work done by a current $c$, when it flows for $t$ seconds through a conductor with a potential-difference $E$ between its ends, is $Ect$. If the current be doing work, such as mechanical work by driving a motor, or chemical work by decomposing some compound, part of the total energy is thus absorbed. If no such work is being done, the whole of the work of the current will appear as heat in the conductor. If $H$ be the quantity of heat developed, measured in thermal units, and $J$ the mechanical equivalent of one thermal unit, then, on the assumption that no other work is done, we have the relation

$$HJ = Ect.$$

But, by Ohm's law, $E$ is equal to $cR$, where $R$ is the resistance of the conductor, at its temperature at the time of the experiment. Thus,

$$HJ = c^2Rt.$$

If we measure the current in electromagnetic units, as, for instance, by the use of a tangent galvanometer, and observe the heat developed in an insulated coil of wire immersed in a liquid in a calorimeter, the first equation, which, as we have seen, is an immediate consequence of the definition of electromotive force, enables us to determine experimentally the electromotive force between the ends of the coil in absolute electromagnetic units. In the same way, the second equation leads to a knowledge of the resistance of the coils, also in absolute electromagnetic units.

On the other hand, if we know these electrical units from other experiments, the equations enable us to find a value for $J$, the mechanical equivalent of the thermal unit. Many experiments have been made for this purpose by Joule, who established the $c^2R$ relation by direct observation, by Griffiths, and by other observers. In the latest experiments by Callendar and Barnes, instead of heating a mass of water in an ordinary calorimeter, the current was passed along a platinum wire stretched along the axis of a glass tube through which flowed a stream of water. Observations were made on the temperature of the water before and after passing the tube, and, when a steady state was reached, a value was thus obtained for the rate of heat development. The electromotive force $E$ between the ends of the wire was measured, and also the electromotive force between the ends of a standard resistance coil of thick wire placed in series with the wire in the calorimeter. The latter observation, with a knowledge of the resistance of the standard coil, which was not appreciably heated by the current, gave the current-strength through the apparatus. Thus the value of $Ect$ was estimated.

In the equation

$$HJ = c^2 Rt$$

the electrical quantities are expressed in absolute electromagnetic units, based on the centimetre, the gramme, and the second. To express our results in practical units, we must remember that the ampere is one-tenth of the absolute unit of current, and the ohm is $10^9$ absolute units of resistance. Thus, writing $C$ for the current in amperes and $R'$ for the resistance in ohms, and putting for $J$ its value $4\cdot18 \times 10^7$ ergs per calorie, we obtain for the number of calories of heat evolved

$$H = 0\cdot239\ C^2 R't.$$

The amount of work done by a current of one ampere flowing for one second through a resistance of one ohm is known as the Joule. It is equivalent to $10^7$ ergs, or absolute C.G.S. units of work.

When this amount of work is being expended per second, the power exerted is known as the Watt.

The $c^2 R$ relation, which sometimes goes by the name of Joule's law, shows us that the heat developed by a current does not depend on the direction of that current. Hence, the heating effect has been employed for the measurement of alternating currents, which cannot be estimated by a tangent galvanometer. In Cardew's voltmeter, for instance, the heating of a long stretched wire increases its length, and moves a hand over a scale which is graduated empirically.

## REFERENCES.

*Elements of the Mathematical Theory of Electricity and Magnetism* ; by Sir J. J. Thomson; Chapter IX.

*Electricity and Magnetism* ; by J Clerk Maxwell ; Chapters VI, VII, VIII, IX, X, XI, XII.

Articles on Electrical Measurements, Galvanometers, etc. in the *Dictionary of Applied Physics.*

# CHAPTER VI

## THERMO-ELECTRICITY

Thermo-electricity.  Application of Thermo-dynamics.  Thermo-electric
diagrams.  Thermo-electric apparatus.

**49.** THE heating effects of currents, due to the resistance of
Thermo-        the material of the conductor, have been considered
electricity.      already.  There is, however, between thermal and electrical phenomena another connexion, which was discovered in 1821
by Seebeck.  He found that, in a closed circuit consisting of two
different metals, if the two junctions were kept at different
temperatures, a permanent current flowed.  Thus, if one junction
of a copper-iron circuit be kept in melting ice and the other in
boiling water, it will be found that a current passes from copper to
iron across the hot junction.  If,
however, the temperature of the hot
junction be raised gradually, the
electromotive force in the circuit
slowly reaches a maximum, then
sinks to zero, and finally is reversed.
The currents thus obtained can, of
course, perform work, and hereafter
we must look for the source of their
energy.

In the year 1834, Peltier discovered that, when a current is passed
across the junction between two
different metals, a reversible evolution or absorption of heat takes place.
This effect may be demonstrated by

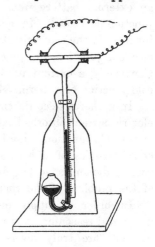

Fig. 74.

fixing the junction, preferably of bismuth and antimony, in an air-thermometer, as shown in Figure 74. The rate of heat-production is proportional to the first power of the current, and thus the total heat developed may be expressed by $\Pi ct$, where $c$ is the strength of the current, $t$ the time, and $\Pi$ the coefficient of the Peltier-effect. Unlike the Joule-effect, which depends on the square of the current, the Peltier-effect is reversible, heat being evolved when the current passes one way across a junction, and absorbed when the current passes in the other direction. With large current-densities, the Peltier-effect is usually small compared with the Joule-effect, but, since the Joule-effect depends on the square of the current, it may be diminished in relative importance by reducing the strength of the current. In ideal conditions, then, when we may imagine that an indefinitely small current passes, the Joule-effect may be neglected compared with the Peltier-effect.

**50.** It is important to notice the relation between the direction of the thermo-electric current and the sign of the Peltier-effect. Seebeck showed that, in a copper-iron circuit at moderate temperatures, the thermo-current passes from copper to iron across the hotter junction. Peltier found that, if a current be forced by means of an external battery from copper to iron across a junction, the junction is cooled. In general, if a current be forced across a junction in the same direction as the thermo-electric current flows at a hot junction, the junction is cooled, that is, heat is absorbed. Conversely, a current passing in the normal direction across the cold junction of a thermo-electric circuit evolves heat.

Application of Thermo-dynamics.

In a thermo-electric circuit, then, the passage of the thermo-electric current absorbs heat at the hot junction, and gives up heat at the cold junction. Lord Kelvin recognised that these were the characteristics of a heat-engine, and applied the principles of thermo-dynamics, so largely his own creation, to the elucidation of the problem of the thermo-electric circuit.

In order to apply quantitatively the results of thermo-dynamical reasoning, it is necessary to be sure that the processes involved are truly reversible. Now, in the present case, two irreversible processes are known to occur: firstly, the frictional heat developed by the current as investigated by Joule, and secondly the conduction of heat along the metals from hotter to colder places. As indicated above, however, by imagining the currents to be restricted to indefinitely small values, the Joule-effect becomes negligible compared with the reversible heat-effects. The conduction of heat may also be ignored if it proceeds independently of the current, except for the reversible Thomson-effect considered below. This independence it is necessary to assume, though the assumption seems reasonable.

Let us, then, treat the thermo-electric circuit as a reversible heat-engine, which, when unit quantity of electricity passes round the circuit, absorbs a quantity $\Pi_1$ of heat at an absolute temperature $T_1$, gives up a quantity $\Pi_2$ of that heat to a refrigerator at an absolute temperature $T_2$, and converts the rest of it into electrical work, which, when unit quantity of electricity passes round the circuit, is measured by $E$, the total electromotive force of the system. By the well-known laws of reversible heat-engines, we now see, firstly, that

$$\frac{\Pi_1}{\Pi_2} = \frac{T_1}{T_2},$$

or

$$\frac{\Pi_1}{T_1} - \frac{\Pi_2}{T_2} = 0,$$

heat evolved, that is leaving the circuit-system, being taken as negative.

Secondly, by the law of the efficiency of a reversible heat-engine, we have

$$\frac{E}{\Pi_1} = \frac{T_1 - T_2}{T_1},$$

or

$$E = \frac{\Pi_1}{T_1}(T_1 - T_2) = \frac{\Pi_2}{T_2}(T_1 - T_2).$$

Thus, on the assumption that the Peltier-effects represent the only reversible heat-changes in the system, we see that, if one junction of a thermo-electric circuit be kept at constant temperature, the total electromotive force round the circuit should increase uniformly with the difference of temperature between the junctions. But, as stated above, in certain circuits, the electromotive force is found to rise to a maximum, decrease and then reverse, as the temperature of one junction is raised continuously. This observation led Lord Kelvin to conclude that the Peltier-effects were not the only reversible heat-changes in a thermo-electric system. He suggested that other reversible heat-effects might be found in the substance of the individual metals, where temperature gradients existed along them, and, by a series of careful experiments, he confirmed this idea experimentally. In a copper bar, heat is carried with the electric current when it flows from hot regions to cold ones; and, on the other hand, when the current flows from cold regions to hot ones, these hot parts of the bar are cooled. In iron these effects are reversed. On the analogy of the flow of a fluid along a channel, we may describe these results by saying that the specific heat of electricity is positive in copper, but negative in iron. If $\sigma$ denote this specific heat of electricity, and $\delta T$ the small difference between the temperatures of two points on a bar, then the heat absorbed, when unit quantity of electricity passes from one point to the other, may be expressed as $\sigma \delta T$, and this relation serves to define the quantity $\sigma$.

The essential difference between the evolution of heat due to this Thomson-effect, and that due to the Joule-effect, should be noted. The frictional heat varies as $c^2 R$, and vanishes in the ideal condition of a very small current. The Thomson-effect varies as the first power of the current, and thus is reversed and becomes a heat-absorption when the current is reversed. The Thomson-effect, then, like the Peltier-effect, is an integral part of the reversible heat-engine which is constituted by a thermo-electric circuit. As the current passes from a region of temperature $T$ to one of temperature $T - \delta T$, heat to the amount of $\sigma \delta T$ is absorbed. Figuratively we may represent the wire as composed of a number of little elements of volume, at the junctions between which occur reversible heat-effects, similar to the Peltier-effects at the junctions between the wires of different metals. Thus, in passing along a wire from one end where the absolute temperature is $T_1$ to the other end where it is $T_2$, an amount of heat is absorbed equal to the sum of all the small quantities $\sigma \delta T$, or, in the notation of the integral calculus, $\int_{T_1}^{T_2} \sigma dT$. Along the other wire of the circuit, where, as we pass with the current, the change of temperature is reversed, the absorption of heat is, in like manner, $\int_{T_2}^{T_1} \sigma' dT$ or $-\int_{T_1}^{T_2} \sigma' dT$.

Instead of a simple heat-engine, with one source and one refrigerator, we have in the thermo-electric circuit a complex engine, with a number of sources from which heat is absorbed, and a number of refrigerators to which heat is given up. Some of these sources and refrigerators are the main junctions, where Peltier-effects occur; others are the hypothetical junctions between the little elements of volume in the individual wires, where, owing to a temperature gradient, Thomson-effects are found. But all these heat-effects are reversible; they all are involved in the working of the thermo-electric engine.

When one unit of electricity passes round the circuit, the work done at each junction, or between each pair of volume elements, is the electrical equivalent of the heat there absorbed, somewhat as the work done by expansion in contact with the hot body in Carnot's theoretical heat-engine, when using an ideal gas as the working substance, is the mechanical equivalent of the heat then absorbed into the cylinder. And, as in Carnot's engine, some of this work is reconverted into heat when the working substance is compressed in contact with the cold body, so, in our thermo-electric engine, part of the electrical energy is reconverted into heat at the other Peltier-junction and at the interfaces, if any, between volume-elements in the wires, where the sign of the Thomson-effect requires an evolution of heat.

Again, in Carnot's engine, according to the principle of the conservation of energy, the balance of useful work is the equivalent of the excess of the heat energy obtained from the source over that given up to the condenser. So here, the balance of electrical energy, obtained per unit electric transfer, is the equivalent of the excess of the heat absorbed at some places in the circuit over that given up at other places. But, this quantity of electrical energy is the effective electromotive force $E$ acting round the circuit, and thus we get

$$E = \Pi_1 - \Pi_2 + \int_{T_1}^{T_2} (\sigma - \sigma') \, dT.$$

Again, as we have shown that, for the simple Peltier-circuit (page 139),

$$\frac{\Pi_1}{T_1} - \frac{\Pi_2}{T_2} = 0,$$

so, for our present more complex case, we may show that

$$\frac{\Pi_1}{T_1} - \frac{\Pi_2}{T_2} + \int_{T_1}^{T_2} \left( \frac{\sigma}{T} - \frac{\sigma'}{T} \right) dT = 0,$$

an equation which expresses the well-known relation of the conservation of entropy in a reversible system.

Now, by differentiating this expression with regard to $T$, that is by finding the value of its rate of change with temperature, we get rid of the integration sign, and may write

$$\frac{d}{dT} \left( \frac{\Pi}{T} \right) + \frac{\sigma - \sigma'}{T} = 0.$$

Thus, 
$$\sigma - \sigma' = T \frac{d}{dT} \left( \frac{\Pi}{T} \right)$$

$$= T \left\{ \frac{1}{T} \frac{d\Pi}{dT} - \frac{\Pi}{T^2} \right\} = \frac{d\Pi}{dT} - \frac{\Pi}{T}.$$

Our equation for $E$ now gives

$$E = \Pi_1 - \Pi_2 + \int_{T_1}^{T_2} \left\{ \frac{d\Pi}{dT} - \frac{\Pi}{T} \right\} dT$$

$$= \int_{T_2}^{T_1} \frac{\Pi}{T} dT$$

Differentiating this expression with regard to $T$, we obtain

$$\frac{dE}{dT} = \frac{\Pi}{T},$$

or 
$$\Pi = T \frac{dE}{dT}.$$

Thus the coefficient of the Peltier-effect appears in this expression as measured by the product of the absolute temperature of the junction and the rate of change with the temperature of that junction of the total electromotive force $E$ round the circuit. Our original equation for $E$, however, on page 141, showed each

Peltier-effect contributing towards the total electromotive force, as a local potential-difference at the junction. These two aspects of the Peltier-effect should be remembered. On the theory here given, the Peltier-effect at a junction is the thermal equivalent per unit electric transfer of the local step of potential at the interface. But it is also measured by the product of the absolute temperature of the junction and the rate of change with the temperature of that junction of the total electromotive force of the circuit (not the local electromotive force at the junction).

**51.** The properties of a thermo-electric circuit may be illustrated by means of a diagram in a manner due to the late Prof. Tait. The diagrams are based on two experimental principles:

*Thermo-electric diagrams.*

(1) The electromotive force round a circuit when the junctions are kept at temperatures $T_1$ and $T_3$ is the sum of the electromotive forces when the junctions are kept first at $T_1$ and $T_2$, and then at $T_2$ and $T_3$.

(2) The electromotive force round a circuit of the two metals $A$ and $C$ is the sum of the electromotive forces round two circuits, one composed of the metals $A$ and $B$, and another composed of the metals $B$ and $C$, the temperature limits being the same for all the circuits.

It was found by Le Roux that in lead the Thomson-effect is inappreciable. Hence it is convenient to take lead as the standard metal, and draw the thermo-electric diagram of another metal with reference to lead. By the second of the experimental results enunciated above, it follows that the relative position of two curves drawn to represent the thermo-electric properties of two metals with reference to lead, represents the thermo-electric properties of those two metals with reference to each other.

The most obvious method of plotting the results of thermo-electric measurement is to take the temperature of one junction as abscissa and the electromotive force round the circuit as ordinate, the temperature of the other junction being kept constant. Curves approximating to parabolas may thus be obtained. A more convenient and usual method, however, consists in taking as abscissae the absolute temperatures, and as ordinates a quantity called the thermo-electric power, which is defined as the electromotive force round the circuit when the junctions are kept half a degree above and below the temperature in question. It is clear that this definition is equivalent to making the thermo-electric power equal to the quantity $dE/dT$. In general, when the diagrams are drawn with reference to the standard substance lead, these curves are found to be straight lines.

In Figure 75, let $OL_1$ be taken as the axis to represent the

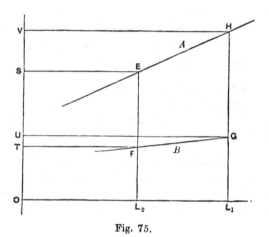

Fig. 75.

standard metal lead. Let $FG$ and $EH$ be the curves for the thermo-electric power of two metals $A$ and $B$ with reference to lead. Then it follows that the thermo-electric power of $A$ with reference to $B$ is represented by the difference-ordinate $FE$ at a temperature of $L_2$ and by $GH$ at a temperature of $L_1$.

Since the length of the ordinate $L_2F$ denotes the electromotive force round a circuit composed of lead and the metal $B$ per degree difference of temperature, it follows that, for a small temperature difference $\delta T$, the electromotive force is represented by the area of a narrow strip of height $L_2F$, and of breadth $\delta T$. Thus, for the considerable temperature-difference $L_2L_1$, the total electromotive force round the circuit is represented by the sum of the areas of the corresponding strips, that is, by the area $L_2FGL_1$. Similarly, for the two metals $A$ and $B$, the electromotive force round the circuit is represented by the area $FEHG$.

Now let us consider the Peltier-effect at the hot junction at the temperature $OL_1$. We have shown on page 139 that the Peltier-effect is equal to $TdE/dT$, and the difference-ordinate $GH$ is the thermo-electric power or $dE/dT$. Thus the electrical equivalent of the Peltier-effect is represented by the area $HGUVH$. Similarly, the Peltier-effect at the cold junction at the temperature $OL_2$ is represented by the area $EFTSE$.

Our expression for the total electromotive force, obtained from the principle of the conservation of energy, namely

$$E = \Pi_1 - \Pi_2 + \int_{T_2}^{T_0} (\sigma - \sigma')\, dT,$$

shows that the electromotive force round the circuit is the algebraic sum of the thermal equivalents of the Peltier- and Thomson-effects. The total electromotive force is represented by the area $HGFEH$, the positive Peltier-effect at the hot junction by the area $HGUVH$, and the negative Peltier-effect at the cold junction by the area $EFTSE$. Now the electromotive force area $HGFEH$ may be made up thus:

Area $HGFEH$ = area $HGUVH$ + area $GFTUG$ − area $EFTSE$
$$\hspace{4cm} - \text{area } HESV.$$

Identifying the areas $HGUVH$ and $EFTSE$ with the Peltier-effects in the equation for $E$, we see that the Thomson-effect in the metal $A$ must be represented by the area $HESVH$, and the Thomson-effect in the metal $B$ by the area $GFTUG$. Thus all the properties of the thermo-electric circuit are represented on the diagram.

Figure 76 shows a thermo-electric diagram for a number of metals. The abscissae represent temperatures measured on the centigrade scale, and the ordinates the thermo-electric powers in microvolts per degree. From what has been said, it will be clear that the thermo-electric behaviour of any pair of metals may be tabulated completely on such a diagram, in a form convenient for future use. It is seen that the Peltier-effect for iron and copper vanishes at a temperature of about $280°$ C. When the hot junction reaches that temperature, called the neutral point, the total electromotive force of a copper-iron circuit, with its cold junction kept at a constant temperature, is a maximum; beyond that temperature the area between the two lines is to be reckoned negative, and subtracted from the area to the left of the neutral point.

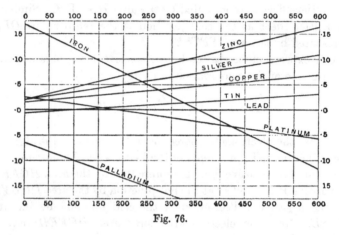

Fig. 76.

**52.** Thermo-electric currents, which are often an inconvenience in delicate measurements of resistance, have been applied with advantage in two ways. A thermo-couple is used sometimes as a practical thermometer, and a thermo-couple, or series of thermo-couples, is employed to detect and measure incident radiation of small intensity.

Thermo-electric apparatus.

An instrument which was long used to detect the incidence of radiation is shown in Figure 77. It is known as the thermopile,

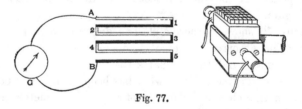

Fig. 77.

and consists of a number of bars of bismuth and antimony soldered together in series, and placed so that all similar junctions, where a current passing round the circuit would flow from bismuth to antimony, are collected together at one face of the pile. If this face becomes slightly hotter than the other, an electromotive force is set up at each junction in the face. These electromotive forces are in the same direction, and will drive a current through a galvanometer placed in the circuit. More modern forms of apparatus consist of thin wires instead of thick bars. The thermal capacity is thus reduced and the sensitivity much increased.

Much greater, however, is the sensitiveness of the radio-micrometer, an instrument invented by Mr Vernon Boys. A single loop of pure copper wire has its ends soldered to a bismuth-antimony junction, to which is attached a small blackened copper disc. When

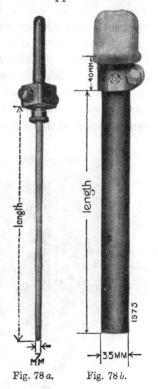

radiation falls on the disc it is heated, and the junction becomes warmer than the rest of the circuit. A current is thus set up round the copper loop. The loop is suspended between the poles of a strong permanent magnet, forming a galvanometer of the moving-coil type. A light mirror attached to the thermo-junction is used to reflect a spot of light on to a scale or the image of a scale into a telescope. It is said that this instrument will detect the radiation from a candle-flame at a distance of more than a mile.

The most important industrial application of thermo-electric currents is the measurement of high temperatures by means of thermo-electric pyrometers. For the thermo-couple, base metals may be used for temperatures up to 1000°C., and platinum with an alloy such as that of platinum and rhodium up to 1400°C. The laboratory type of such a couple is shown in Fig. 78 a and a form suited to the factory in Fig. 78 b.

Fig. 78 a.          Fig. 78 b.

The galvanometer is of the moving coil type with a high resistance (Fig. 79), so that its deflection depends on the E.M.F.

10—2

generated. It can be graduated directly in temperatures. The
cold junction should be kept at a constant temperature, *e.g.* 0° C.,
but, if this cannot be done, the zero of
the galvanometer scale can be set to
correct for any variation therefrom.

For research and for industrial
work, it is often necessary to obtain
an automatic and permanent record
of changing temperatures. With a
resistance thermometer, this can be
done by means of Callendar's Recorder,
which is a self-adjusting Wheatstone's
bridge. But a simpler instrument
applicable either to the resistance or
the thermo-electric thermometer is
the thread recorder. This consists of a
galvanometer with either a suspended

Fig. 79.

or a pivotted coil, the pointer of which is automatically depressed
by clock-work every minute or half-minute on to a chart covering a
drum. A thread impregnated with ink passes between the pointer

Fig. 80 *a*.

and the chart, and when the pointer is depressed it forces the thread on to the paper and so gives a dot to record the temperature. As the drum revolves slowly the successive dots merge into a line. One of these recorders, made by the Cambridge and Paul Instrument Company, is shown in Fig. 80 a, and a record in Fig. 80 b.

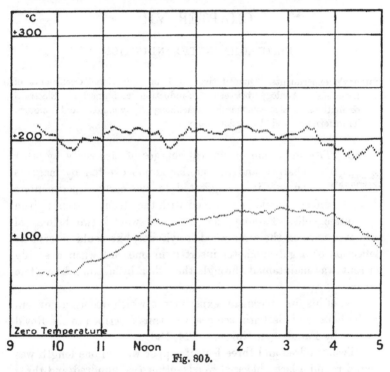

Fig. 80 b.

For temperatures above 1400°, a radiation pyrometer must be employed. The radiation from a small opening into a furnace, or better from a tube with its closed end inserted into the furnace, is focused on to a thermo-couple, and the deflection of a galvanometer can then be used as a measure of the intensity of the radiation, which depends on the fourth power of the temperature.

### REFERENCE.

Articles on Thermo-couples and Thermodynamics in the *Dictionary of Applied Physics*.

# CHAPTER VII

## ELECTROMAGNETIC INDUCTION

Faraday's experiments. Quantitative Laws of Induction. Coefficients of induction. Analogy between self-induction and inertia. Electrical oscillations. Electromagnetic machinery. Dynamos and motors. Transformers and Induction Coils.

**53.** THE induction of statical charges of electricity by other
Faraday's charges, and the similar action exerted by magnets
experiments. on soft iron, suggested to the early experimenters
that like effects should be obtained with the steady currents given
by voltaic cells. Faraday, for instance, wound two helices of
insulated wire on the same wooden cylinder, but could observe no
deflection of a galvanometer inserted in one coil when a steady
current was maintained through the other by a powerful voltaic
battery.

One of his first successful experiments, which opened a new era
in the history of electrical science, was thus described to the Royal
Society by Faraday on November 24, 1831.

"Two hundred and three feet of copper wire in one length was
wound round a large block of wood; other two hundred and three
feet of similar wire were interposed as a spiral between the turns
of the first coil, and metallic contact everywhere prevented by
twine. One of these helices was connected with a galvanometer,
and the other with a battery of one hundred pairs of plates four
inches square, with double coppers, and well charged. When the
contact was made, there was a sudden and very slight effect at the
galvanometer, and there was also a similar slight effect when
the contact with the battery was broken. But while the voltaic
current was continuing to pass through the one helix, no galvano-
metrical appearances nor any effect like induction upon the other

helix could be perceived, although the active power of the battery was proved to be great, by its heating the whole of its helix, and by the brilliancy of the discharge when made through charcoal.

"Repetition of the experiment with a battery of one hundred and twenty pairs of plates produced no other effect; but it was ascertained, both at this and the former time, that the slight deflection of the needle occurring at the moment of completing the connexion, was always in one direction, and that the equally slight deflection produced when the contact was broken, was in the other direction.

"The results which I had by this time obtained with magnets led me to believe that the battery current through one wire, did, in reality, induce a similar current through the other wire, but that it continued for an instant only, and partook more of the nature of an electrical wave passed through from the shock of a common Leyden jar than of the current from a voltaic battery, and therefore might magnetize a steel needle, though it scarcely affected the galvanometer.

"This expectation was confirmed; for on substituting a small hollow helix, formed round a glass tube, for the galvanometer, introducing a steel needle, making contact as before between the battery and the inducing wire, and then removing the needle before the battery contact was broken, it was found magnetized.

"When the battery contact was first made, then an unmagnetized needle introduced into a small indicating helix, and lastly the battery contact broken, the needle was found magnetized to an equal degree apparently as before; but the poles were of a contrary kind."

With the much more delicate galvanometers we now possess, it is easy to repeat Faraday's experiments with the primary current derived from a single voltaic cell, and to show that similar transient currents are produced by moving the primary and secondary circuits relatively to each other, or by moving a permanent magnet relatively to a coil connected with a galvanometer.

In all such cases it will be seen that the number of tubes of magnetic induction passing through the secondary circuit is altered, and, in general, we may say that a current is induced in the same direction as the primary current (or in the same direction as the current equivalent to the inducing magnet) whenever the number of tubes of magnetic induction threading the secondary circuit is diminished, and in the contrary direction when the number of magnetic tubes is increased. The positive direction of the tubes of induction through the secondary circuit is taken to be that in which they pass when they are caused by a positive current in the secondary circuit itself.

Since currents flowing in the same direction attract each other, it follows that, when a circuit carrying a current is moved nearer to a second circuit, the inverse current induced therein causes the two circuits to repel each other. When the primary recedes, the two similar currents attract each other. Thus, in each case, the induced current is in that direction which tends to stop the motion producing it. This result, which is known as Lenz's law, is an example of the universal physical principle whereby any external change acting on a system produces within the system a change which tends to resist the effects of the external change. In the form of Lenz's law, the principle is often of use in enabling us to see at a glance the direction of an induction current.

Faraday's experiments led him at once to an explanation of the action of the apparatus known as Arago's disc. If a copper disc be rotated in its own horizontal plane beneath a compass-needle, which is shielded from the air draughts of the disc by an interposed glass screen, the needle will be found to be dragged round with the disc. Explanations of this phenomenon had been offered, referring it to "induced magnetism" in the copper disc. Faraday saw that, by motion in the magnetic field of the needle, secondary eddy currents were set flowing in closed curves in the substance of the disc. These currents will be in directions such that they tend to stop the relative motion of the needle and disc. The needle, therefore, follows the disc as it rotates.

The current induced in one coil by starting or stopping a current in another is much greater if the coils be wound on an iron core. This suggests that it is the change in the magnetic induction, and not that in the magnetic force, which is involved.

By winding a primary coil over an iron ring, and using as a secondary a few turns of wire surrounding both primary and iron, it is possible to investigate the magnetic properties of the iron, and to obtain hysteresis curves similar to those given in § 33, Chapter IV. In many cases this is the most convenient method to adopt.

**54.** Faraday showed that the induced currents were greater in proportion as the rate of change of the magnetic field *Quantitative laws of induction.* was increased, but the exact quantitative magnitude of the induced electromotive force was first given by F. E. Neumann in 1845.

We may deduce by dynamical principles the laws of induction of currents from the known mechanical forces on currents placed in magnetic fields.    Ampère's formula for the mechanical force $F$ on an element of current $c$ of length $\delta l$, when placed in a magnetic field at an angle $\theta$ with the direction of the magnetic induction $B$, is (p. 108)

$$F = c\,\delta l\,B \sin\theta,$$

the force acting at right angles both to the current and to the magnetic induction.

Let us imagine that the element $ab$ of the circuit in Figure 81 moves with constant velocity through a distance $\delta x$ at right angles to itself in the direction of the force.    To maintain the continuity of the circuit we may suppose that $ab$ slides along parallel rails, or that its ends float along mercury troughs.

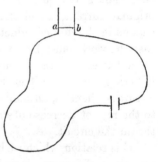

The work done by the moving element is measured by the force into the distance or

$$cB \sin\theta\,\delta l\,\delta x.$$

Now $\delta l\,\delta x$ is the area between the

Fig. 81.

rails moved over by the element; that is, it is the increase of area of the circuit.    The magnetic induction $B$ denotes the number of tubes of induction per unit area normal to their direction.    Thus $B \sin\theta\,\delta l\,\delta x$ is equal to $\delta N$, the increase in the number of tubes of induction through the circuit produced by the movement of the current-element $ab$, the positive direction of the tubes of induction being that of tubes due to a positive current in the circuit. The work done during the movement may be written as $c\delta N$.

Now let us suppose that the current in the circuit is maintained by a battery with an electromotive force $E$. During the time $\delta t$ of the movement of the element $ab$, the battery will do a quantity of work equal to $Ec\delta t$. By the principle of the conservation of energy, part of this energy is used in doing work $c\delta N$ in moving the element $ab$ against the magnetic forces, and part is expended in the circuit as a quantity of heat $c^2 R\delta t$. If there be no change in the energy of the magnetic field, we have, therefore,

$$Ec\delta t = c^2 R\delta t + c\delta N,$$

or
$$c = \frac{E - dN/dt}{R}.$$

Hence there is an additional electromotive force in the circuit equal to the rate of decrease of the number of tubes of induction which thread the circuit. We may imagine that the original current and electromotive force are made as small as we please— the result still holds. It follows that, even in a circuit with no original current, an electromotive force must be induced by a change in the flux of induction through the circuit, though in this case the work must be done by some external agency, instead of by a battery in the circuit. Thus, in general, we obtain the following result:—When the magnetic induction through any circuit is changed, an induced electromotive force is set up, equal to the rate of decrease of the number of lines of induction which thread the circuit.

This relation, which we have deduced from the known magnetic properties of currents, may be verified or established by direct experiment. Two coils of wire are placed near each other, as shown in Figure 82. In circuit with one coil is connected a battery, a resistance box, a tangent galvanometer, and a reversing key. In circuit with the other coil is placed a second resistance box and a ballistic galvanometer (see page 113). When a steady current passes through the first or primary coil, there is no current in the secondary, but, if the primary current be started suddenly, stopped or reversed, a throw of the ballistic galvanometer indicates a rush of current through the secondary circuit.

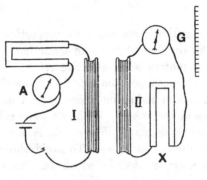

Fig. 82.

The primary current produces a magnetic field, and some of the tubes of magnetic induction due to it pass through the secondary coil. If the number of these tubes be altered as described above, a transient current will be observed in the secondary circuit. Suppose, for instance, that we reverse the primary current. Then all the tubes of induction, which at first threaded the secondary coil in one direction, will be reversed suddenly, so that, if $N$ be the number of tubes originally passing through the secondary coil, the number finally passing is $-N$, and the sudden change in the number is $2N$. The magnetic field, and therefore $N$, is proportional to the strength of the primary current; thus any required change in $N$ can be obtained by altering the strength of the current which is to be reversed; the current used can be measured by the tangent galvanometer. By such experiments we can show that the total quantity of electricity passing in the secondary current is proportional to the total change in the number of tubes of magnetic induction threading the secondary circuit, and inversely proportional to the total resistance of the secondary circuit. Therefore, if, owing to a change $\delta N$ in the number of tubes of induction through the circuit, a current $c$ pass in a circuit of resistance $R$ for a small time $\delta t$, we may write

$$c\,\delta t \propto \frac{\delta N}{R}.$$

By Ohm's law, the product of $c$ and $R$ is the electromotive force $E$ in the circuit, thus

$$E \propto \frac{\delta N}{\delta t},$$

or the electromotive force must be proportional to the rate of change in the number of tubes of induction.

In order to obtain a quantitative relation, we must use two coils placed in such a position that it is possible to calculate the number of tubes of induction which thread one coil when unit current passes in the other. With two large coils placed parallel to each other as shown in Figure 82, it would be difficult to calculate the number of tubes. To avoid this difficulty, a form of apparatus may be used in which the magnetic field of the primary current is uniform. For instance, the field due to a current $c$ in a large circular coil with $n_1$ turns of wire of radius $r_1$ is nearly uniform near the centre, and equal to $2\pi n_1 c/r_1$ (page 110). In air, this expression represents the number of tubes of induction per unit area normal to their direction; and the number of tubes through a small coil of $n_2$ turns with radius $r_2$ placed at the centre of the large coil with the planes of the two coils coincident is $\frac{2\pi n_1}{r_1} \pi n_2 r_2^2$, or $\frac{2\pi^2 n_1 n_2 r_2^2}{r_1}$, when unit current flows round the large coil.

Again, the field within a long solenoid is $4\pi nc$, where $n$ is the number of turns of wire per unit length. When placed coaxially within the solenoid, a small circular coil of $n_2$ turns with radius $r_2$ will be threaded by a number of tubes of induction equal to $4\pi n . \pi n_2 r_2^2$ or $4\pi^2 n n_2 r_2^2$ when unit current passes through the solenoid. By either of such pieces of apparatus the laws of induction may be verified quantitatively by the use of formulæ obtained in this book. The result of the experiments shows that the total electric transfer round the circuit is, when expressed in absolute electromagnetic units, with proper attention to signs, equal to the total decrease in the number of tubes of induction divided by the resistance of the circuit. Thus, as before,

$$c\delta t = -\frac{\delta N}{R}, \quad \text{and} \quad E = -\frac{dN}{dt}.$$

It should be noted that this induced electromotive force acts in the primary coil as well as in the secondary, for all its own lines of induction must pass through the primary circuit. On making a current in a coil, a transient electromotive force acts in the negative direction, and the current is prevented from reaching instantaneously its full value. So also on breaking the circuit, the positive induced electromotive force tends to prolong the current. In either case, the result is that no abrupt change can occur in the total number of tubes of induction which thread the circuit.

It should be noticed that, with two circuits, the induced secondary current will itself induce a tertiary electromotive force in the primary coil. When a primary current in what we will call the positive direction is started, a secondary current is produced in the negative direction. The starting of this will re-induce an electromotive force in the positive direction in the primary. It will thus help the primary current to start, and to increase towards its maximum value. The presence of a second circuit then will diminish the effect of self-induction in any circuit; it will enable an applied electromotive force sooner to establish its current, and make it easier for that current to cease when the electromotive force is removed. This effect is of great practical importance.

**55.** The number of tubes of magnetic induction which thread one circuit when unit current is flowing round another is called the coefficient of mutual induction between the two circuits. The importance of this coefficient will be clear if we notice that the total number of tubes threading the one circuit, when a current $c$ passes through the other, is $Mc$, where $M$ is the coefficient of mutual induction. Thus the induced electromotive force in the circuit, $-dN/dt$, is equal to $-Mdc/dt$. Hence, if we know the coefficient of induction and the rate of change of the primary current, we can find at once the secondary electromotive force.

*Coefficients of induction.*

A corresponding coefficient of self-induction $L$ gives us the induced electromotive force in the primary coil itself as $-Ldc/dt$. When the area of a coil is large compared with the dimensions of the wire, the self-induction of the coil may be defined as the number of tubes of magnetic induction which thread it when unit current flows; but this definition is not of general application.

We have seen already that it is possible to calculate the coefficient of mutual induction in simple cases. An arrangement of practical importance consists of a closed iron ring wound over by two coils of wire. Let $r_1$ and $n_1$ denote respectively the radius and the number of turns of wire per unit length of the inner coil, and $r_2$ and $n_2$ the corresponding quantities for the outer coil. If unit current flow round the inner coil, the field of magnetic induction within it is $4\pi n_1 \mu$, which measures the number of tubes per unit area, $\mu$ denoting the permeability of the iron. The total area of the inner coil is $\pi r_1^2$, and the total magnetic flux through the inner coil is $4\pi^2 n_1 r_1^2 \mu$. All these tubes of induction must pass through each turn of wire of the outer solenoid, which surrounds the inner solenoid. The number of such turns is $n_2 l$, and therefore the coefficient of mutual induction is $4\pi^2 l n_1 n_2 r_1^2 \mu$. Now let us suppose that the current is sent through the outer coil. When the current is unity, the magnetic field within is $4\pi n_2$, and the number of tubes of induction which pass through the total area (counting that of each turn of wire) of the inner coil, which only fills part of the area of cross-section of the outer coil, is $4\pi n_2 n_1 l \pi r_1^2 \mu$ or $4\pi^2 l n_1 n_2 r_1^2 \mu$. It follows that the number of tubes of induction passing through the outer coil when unit current flows through the inner is equal to the number which passes through the inner when unit current flows round the outer. This justification of the term "mutual induction" will be shown later to apply to every case.

From what has been said, it follows that the coefficient of self-induction of a single coil wound over a closed iron ring is $4\pi n \cdot n l \pi r^2$ or $4\pi^2 l n^2 r^2$, while that for a long solenoid has approximately the same value.

For a small circular coil with $n_2$ turns of radius $r_2$ placed at the centre of a large circular coil with $n_1$ turns of radius $r_1$, the planes of the coils being parallel, the coefficient of mutual induction must, by the last section, be $2\pi^2 n_1 n_2 r_2^2 / r_1$.

Other simple cases may be solved by the aid of more complicated mathematical analysis, but it is often only possible to determine coefficients of induction by direct experiment. The apparatus shown in Figure 82 on page 155 may be used, and the number of tubes of magnetic induction threading the secondary coil when unit current passes round the primary may be calculated from a measurement of the total electric transfer round the secondary when a known current in the primary is reversed. Here we assume the laws of induction which, on page 155, we used the apparatus to verify.

We have shown above, for one particular case, that the coefficient of mutual induction between two coils is the same whichever coil is used as primary. It is possible to establish this result in a general form by a method which brings out other points of interest.

Let us calculate the mutual potential energy of two currents, by supposing them to be replaced by their equivalent magnetic shells.

The potential at a point due to a magnetic shell of strength $S_1$ is, as we saw on page 102, equal to $S_1 \Omega$, where $\Omega$ is the solid angle which the shell subtends at the point. The work done in bringing up from infinity a pole of strength $m$ to the point is $S_1 \Omega m$.

The second shell, of strength $S_2$ and thickness $\delta x$, may be resolved into a number of minute magnets, each of length $\delta x$ and area of cross-section $\delta a$. Now, the strength of a shell is the magnetic moment per unit area, or $\dfrac{m \delta x}{\delta a}$, where $m$ is the pole strength of the elementary magnet. Therefore $m$ is $S_2 \dfrac{\delta a}{\delta x}$.

The potential energy of the first shell and one pole of the elementary magnet in the second shell is

$$S_1 S_2 \frac{\delta a}{\delta x} \Omega.$$

Now, as we pass from one end of the elementary magnet of the second shell to the other, the solid angle subtended by the first shell changes. Its rate of change as we pass along the axis of the elementary magnet of the second shell away from the first is $\dfrac{d\Omega}{dx}$, and the total change is $\dfrac{d\Omega}{dx} \delta x$. Thus the new value of the solid angle is $\Omega - \dfrac{d\Omega}{dx} \, \delta x$. The pole at the other end of the elementary magnet has a strength $- m$ or $- S_2 \dfrac{\delta \alpha}{\delta x}$, and its mutual potential energy with the first shell is the work done in bringing it to its proper position from an infinite distance, or

$$- S_1 S_2 \frac{\delta \alpha}{\delta x} \left( \Omega_1 - \frac{d\Omega}{dx} \, \delta x \right).$$

For both poles of the element of the second shell the work done is

$$S_1 S_2 \frac{\delta \alpha}{\delta x} \cdot \frac{d\Omega}{dx} \, \delta x = S_1 S_2 \frac{d\Omega}{dx} \delta \alpha.$$

To extend the result to the whole of the second shell, we must integrate this expression over the whole area, and write

$$S_1 S_2 \int \frac{d\Omega}{dx} \, d\alpha.$$

But, by principles similar to those used on page 27, it follows that the magnetic force in any direction is measured by the rate of fall of magnetic potential in that direction. Hence, the normal force at any point on the second shell due to the first shell is $- S_1 \dfrac{d\Omega}{dx}$. This measures the number of tubes of magnetic force passing through unit area, and through an area $\delta \alpha$ the number will be $- S_1 \dfrac{d\Omega}{dx} \delta \alpha$. Therefore, the total number $N_1$ of tubes of magnetic force through the whole area of the second shell due to the first shell will be

$$N_1 = - S_1 \int \frac{d\Omega}{dx} \, d\alpha,$$

$N_1$ being counted positive when the lines thread the second circuit in the same direction as the lines of force of that circuit itself.

Hence, the work done in bringing the second shell to its position against the magnetic forces of the first, that is, the mutual potential energy of the two shells, is $- S_2 N_1$, the product of the strength of the second shell and the magnetic flux passing through it from the first shell.

By calculating in a similar manner the work done in bringing up the first shell while the second is fixed, we obtain, as a new expression for the same mutual potential energy, $- S_1 N_2$, and hence

$$S_1 N_2 = S_2 N_1.$$

Finally, let us replace the shells by two currents of equivalent strengths $c_1$ and $c_2$. We have

$$c_1 N_2 = c_2 N_1.$$

If unit current be flowing in one circuit, the magnetic flux through the other is defined as the coefficient of mutual induction; and since, when both currents are of unit strength, $N_2$ is equal to $N_1$, the fluxes are equal, that is, the number of tubes of magnetic induction passing through the second circuit when unit current flows round the first, is equal to the number of tubes passing through the first when unit current flows round the second.

We see also that the coefficient of mutual induction between two circuits measures their mutual potential energy when unit current circulates round each.

**56.** Since the induced electromotive force in a coil through which passes a varying current is in the opposite direction to the current when it is increasing, and in the same direction as the current when it is diminishing, the effect of self-induction is to tend to check the change in current-strength. Now the inertia of a dynamical system tends to retard its being set in motion when at rest, and to keep it in motion when once started. Thus, just as inertia tends to prevent a change in velocity, so self-induction tends to prevent a change in current-strength.

*Analogy between self-induction and inertia.*

If the self-induction of a coil be very high, as it is in coils of many turns of wire, especially if wound on iron cores, the effect of self-induction will be so great that a rapidly alternating electromotive force will produce hardly an appreciable current. A rapidly alternating current, then, will pass along a straight short path, even if of high resistance, rather than through a low resistance circuit of high inductance, through which a steady current passes readily. A coil which, in this way, is nearly impassable to rapidly alternating currents while allowing direct currents to flow through it readily is often of practical use, and is called a choking coil.

The quantitative investigation of this effect needs the solution of a differential equation; but, even if the proof cannot be followed, the result is of great interest. Let an external electromotive force, which varies harmonically with the time $t$ and is represented by $E \cos pt$, be applied to a circuit with a coefficient of self-induction $L$ and a resistance $R$. This electromotive force makes $p/2\pi$ complete alternations per second, changing direction $p/\pi$ times per second. If $c$ be the current through the coil, the primary electromotive force is $Rc$, while the secondary is $L\,dc/dt$. Thus

$$L \frac{dc}{dt} + Rc = E \cos pt.$$

The solution of this equation is

$$c = \frac{E \cos (pt - \alpha)}{\{R^2 + L^2 p^2\}^{\frac{1}{2}}},$$

where $\tan \alpha = Lp/R$.

The maximum value of the electromotive force is $E$, while that of the current is $E/\{R^2 + L^2 p^2\}^{\frac{1}{2}}$. Now, if a steady electromotive force $E$ were acting, the steady current would be $E/R$. Thus the inertia of the circuit, due to its self-induction, makes the effective resistance greater for alternating currents than for steady currents, and equal to $\{R^2 + L^2 p^2\}^{\frac{1}{2}}$. This effective resistance is called the impedance of the circuit.

The equation for the current shows that the phase of the current lags behind that of the electromotive force. When the alternations are so rapid that $Lp$ is large compared with $R$, $\tan \alpha$ is very large, and hence $\alpha$ approaches a right angle or $\pi/2$. The current in the coil will now be greatest when the electromotive force is zero, and will vanish when the electromotive force is a maximum. Allowance must often be made for this lag of the current in electromagnetic machinery.

The subject of alternating currents has now become of great industrial importance, and many theoretical investigations have been made thereon. From the experimental side, much light has been thrown on the phenomena by the use of an instrument known as the oscillograph, which, first described by M. Blondel, is, in its present form, the invention of Mr Duddell. In its essence, this instrument is a galvanometer of the moving-coil type, the moment of inertia of the coil being so much reduced, and the restoring couple so much increased, that the natural period of vibration is reduced to the 1/5000 or the 1/10000 of a second. By the friction of a viscous oil the motion of the coil is made "dead-beat," that is, the system moves at once to its position of equilibrium, and no series of swings follows a displacement. In consequence of these arrangements, the apparatus will follow the alternations of a

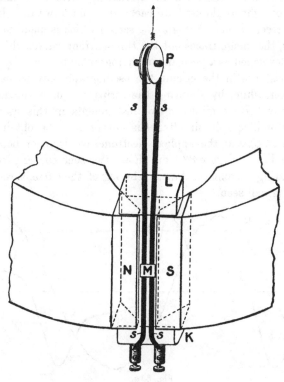

Fig. 83.

current with a frequency of 100 or more per second. The apparatus is represented diagrammatically in Fig. 83. The coil is replaced by a loop of two delicate phosphor-bronze strips, *ss*, arranged parallel to each other and stretched by a light spring so as to lie vertically between the poles of a powerful permanent or electro-magnet. The strips run through minute gaps in the magnetic circuit, and these

11—2

gaps are filled with viscous oil. When a current passes round the loop, one strip moves forward and the other back, just as the coil of a d'Arsonval galvanometer turns through an angle. The relative movement of the two strips is made visible by fixing to them a very small and light mirror $M$, so that any relative shift rotates the mirror about a vertical axis. The horizontal displacement of a spot of light may thus be made to measure the instantaneous value of the current. With an alternating current the image of this spot of light would oscillate backwards and forwards. If this image be received on a photographic screen which is made to move vertically, the image traces out the time-current curve, which can then be developed as a permanent photographic record. By using a high resistance in the circuit, the oscillograph may be used as a voltmeter, while, by shunting the strips, it can be made into an ammeter. By combining two instruments in this way into a double oscillograph, simultaneous curves may be obtained of the current $C$, and the applied electromotive force or potential difference $V$. In Figure 83 $a$ are shown two such curves, given by an alternating current machine. The lag of the current, referred to above, is well seen.

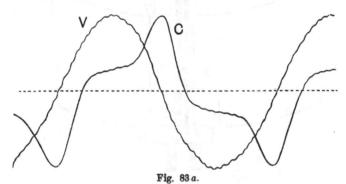

Fig. 83 a.

We have supposed that a periodic electromotive force may be represented by a simple harmonic function of the time—that it varies as the cosine of an angle proportional to the time. Such a curve would somewhat resemble the curve marked $V$ in Figure 83 $a$. Often, however, the actual variations of current or potential-difference are much more complicated, and peaked curves, such as the curve $C$ in the figure, are common. For some purposes—arc lighting for instance—a flat topped wave-form is most efficient, while in other cases—such as that of transformers— a peaked curve has advantages. Hence arises the importance of the oscillograph, which enables us to follow the wave-form, and actually trace out its curve.

In the chapter on the properties of the dielectric medium, we learnt that the path of the energy by which a current is maintained lies through the dielectric medium. The line of the current represents merely the path along which that energy is degraded into heat by something analogous to a frictional slip. Hence, when the energy is alternating in character, successive currents spread into the conductor—spread in with a finite velocity, just as heat is conducted into the ground under the alternating temperatures of day and night, summer and winter. Moreover, just as the quick alternations due to day and night become insensible at a small distance from the surface of the ground, while the slower alternations due to the varying seasons penetrate deeper, so a quickly alternating current will only reach a small distance into the substance of a conductor. This limitation of an alternating current to the outer layers of the conductor may be regarded as an explanation of the increased effective resistance of the conductor when a continuous is replaced by an alternating current.

The quantitative theory of the subject may be investigated by observing that the self-induction of a current flowing along the outer strata of a conductor is less than that of a current flowing within. The magnetic force inside a cylindrical tube carrying a current may be shown to vanish. Hence currents flowing down the outside of a wire produce no magnetic field within, and no induced currents. Thus the inertia of the current will be less if the current flows through the outside layers only.

The analogy between self-induction and inertia may be pushed farther than we have yet indicated. The dynamical inertia of a body, which measures its mass $m$, tends to prevent any change in the velocity $v$. The self-induction $L$ of a circuit tends to prevent any change in the strength $c$ of the current.   Hence work must be done in starting or stopping a current, just as it is in starting or stopping a moving body.   The work in the latter case is equal to the kinetic energy $\frac{1}{2}mv^2$.   Thus, on the analogy we are considering, an amount of work $\frac{1}{2}Lc^2$ is needed to start or stop a current $c$ in a circuit possessing self-induction $L$, and the current, when flowing, must consequently possess an amount of kinetic energy equal to $\frac{1}{2}Lc^2$.

A still more complete analogy may be traced by writing down the equations of motion of a body with a velocity accelerating under the action of a force $F$, and retarded by a frictional resistance which is proportional to the velocity.   If $v$ be the velocity, and $R$ the resistance for unit velocity, we have

$$F = m\frac{dv}{dt} + Rv.$$

Now let us compare this equation with that for an electromotive force $E$ applied to a circuit of resistance $R$ and self-induction $L$. If $c$ be the current,

$$E = L\frac{dc}{dt} + Rc.$$

Thus the analogy appears to be complete.

**56 $a$.**   Important results follow from the analogy between self-induction and inertia.   If a mass $m$ hanging by a spiral spring be pulled down from its position of equilibrium and let go, it performs a series of vertical oscillations, till the resistance to its motion brings it to rest.   If the resistance be small, the oscillations will continue for some time; they will be isochronous, and have a periodic time of $2\pi\sqrt{m/F_1}$, where $F_1$ denotes the resultant force of restitution exerted when the displacement of the body is unity.   Now, as the body oscillates from the middle of its path to one of the ends, the energy changes from the kinetic form to the potential.   If $x$ be the final displacement, the final force acting on the body is $F_1x$; the mean force throughout the extension is $\frac{1}{2}F_1x$, and the work done, or the final potential energy of the body, is $\frac{1}{2}F_1x^2$.

Now let us consider the analogous case of two parallel plates charged with opposite kinds of electricity.   If connexion be

*Electrical oscillations*

made between the plates by means of a conducting wire, an electromotive force acts along it, and a current is set up. If the circuit of the wire possesses appreciable self-induction, the current will possess inertia, and, unless the frictional resistance is very great, the current will tend to flow on, after the condition of equilibrium is reached. The plates will thus acquire reversed charges; the electromotive force will act in the other direction; and electric oscillations will result.

The potential energy of the charged plates is $\frac{1}{2}eV$, where $V$ is the difference of potential, and $e$ the charge on one plate. If $C$ be the electrostatic capacity of the system, $C = e/V$, and the potential energy is $\frac{1}{2}e^2/C$. Now on the analogy of the oscillating spring, the displacement of electricity from its condition of equilibrium, that is the charge on the plate, corresponds with the displacement from its position of equilibrium of the body hanging by the spring. Corresponding with $F_1$, the coefficient of $\frac{1}{2}x^2$ in the expression for the potential energy of the body, we have $1/C$, the coefficient of $\frac{1}{2}e^2$ in the expression for the potential energy of the charged plates.

We may now use our analogy to write down the expression for the period of oscillation of the electric system. The self-induction $L$ of the circuit is analogous to $m$, and $1/C$ to $F_1$. The period of the spring is $2\pi \sqrt{m/F_1}$, and the corresponding expression is $2\pi \sqrt{LC}$.

If the resistance exceed a certain limit, no oscillations will arise. The charges will neutralize each other by means of a steady current, just as the body on the end of the stretched spring will creep slowly to its position of equilibrium if the whole system be immersed in some very viscous liquid.

The interest of these electric oscillations is very great. The discharge from a Leyden jar, the lightning flash itself, will, it is evident, in the right circumstances, produce, not a direct current, but a series of electric oscillations. The frequency of such oscillations will be very great, perhaps rising to hundreds of thousands per second, and rapidly alternating currents, as we have seen, show preferences for different paths very unlike the preferences shown by steady currents. A lightning flash or a discharge from a jar may take a short cut through air rather than go round a metallic circuit of appreciable self-induction. Hence the necessity of avoiding bends and turns in lightning rods.

The oscillations in the spark from a Leyden jar were first verified by direct experiment by Feddersen in 1857, who analysed the spark by viewing it in a mirror rotating very rapidly. The image was drawn out into a band with bright and dark spaces. On placing a high resistance in circuit, the alternations disappeared, and the image became a band gradually fading away at one end.

The periodic character of the discharge from a Leyden jar may also be demonstrated by another beautiful experiment, due to Professor Sir J. J. Thomson. The discharge is made to pass round a coil of a few turns of insulated wire within which is placed a glass bulb containing air at a low pressure. The alternating currents in the coil induce high electromotive forces in the exhausted air, just as they would do in a circuit of wire in the neighbourhood. By this means the resistance of the air is broken down, and a circular current passes in the form of a luminous ring discharge within the bulb. The method has been of service in eliminating the effects of the electrodes in the spectroscopic study of electric discharge through gases.

**57.** Faraday's discovery of electromagnetic induction has proved to be the foundation of a vast industrial development: almost all electrical machinery of practical importance depends on the induction of currents.

Electro-
magnetic
machinery.

An analogy, of great use in designing electromagnetic machinery, may be traced between the permeability of iron and the specific conductivity of an electric conductor. If a bar of iron of great length and uniform cross-section $A$ be placed in a uniform field of magnetic force of intensity $H$, the total number of lines of induction through the iron is $HA\mu$. This quantity, known as the magnetic flux (see p. 78), may be written as $N$.

Now
$$N = \frac{Hl}{\dfrac{l}{A\mu}},$$

and we may compare this equation with that expressing the strength of an electric current $c$, which is equal to the electromotive force $E$ divided by the resistance $R$ of the conductor, or

$$c = \frac{E}{R}.$$

In these two equations, the quantity $Hl$ corresponds with the electromotive force, and may be called the magnetomotive force; the quantity $l/A\mu$ corresponds with the electric resistance, and may be called the magnetic reluctance.

The electric resistance is equal to the specific resistance of the material into the length of the conductor divided by the area of cross-section, and, if the reciprocal of the specific resistance, or the specific conductivity, be denoted by $y$, we have for the total resistance $l/Ay$. Thus $\mu$ is analogous to $y$, except in so far as $\mu$ is not constant, but depends on the magnetomotive force.

The total resistance of an electric circuit is the sum of the individual resistances of its parts, and the total reluctance of a magnetic circuit is the sum of the reluctances of the different parts, whether iron or air, which together complete the circuit.

When the approximate value of $\mu$ which corresponds with the magnetic fields to be used in a machine is known, this law of the magnetic circuit enables us to calculate the number of "ampere-turns" needed in winding and exciting the electro-magnets.

**58.** In dynamos, coils of wire wound on iron shuttles or rings
Dynamos and    are rotated in the fields of strong electro-magnets,
motors.            excited either separately by an independent machine,
or by the current they themselves induce in the rotating coils.

The principle of the dynamo is best explained by a diagram such as Figure 84. Let $CC'$ be a circular coil of wire, spinning in a magnetic field where the lines of induction run in the direction of the arrow $AB$. When the magnetic field is that of the earth, the apparatus is known as an earth-coil or earth-inductor. When the plane of the coil lies along $AB$, no lines of induction pass through it. As the coil spins clockwise through 90°, the number of lines threading it continually increases and reaches

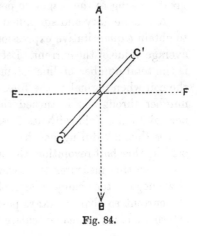

Fig. 84.

a maximum at *EF*. From *EF* onwards through 90°, the number decreases. Consequently, the induced electromotive force is reversed as the coil passes through the position *EF*, at right angles to the lines of induction. When the coil again lies along *AB*, no lines thread it, and, as it rotates through the third quadrant, the number again increases. But now they enter the other face of the coil, and must be counted as negative. The present increase, then, produces an electromotive force in the same sense as did the decrease in the second quadrant, and no change of sign occurs in the electromotive force when the coil passes the line *AB*, that is, when its plane lies along the direction of the lines of induction. The negative lines, however, will begin to decrease when *EF* is again passed. Thus the electromotive force in the coil changes sign whenever the coil passes the position where its plane is normal to the magnetic field.

By connecting the ends of the coil with brass pieces on an insulating drum which rotates with the coil, the currents may be led away through wire brushes to the external circuit. By proper arrangements, either alternating currents may be obtained, or the alternating currents in the coil may be commutated and passed always in the same direction round the external circuit. In practice, many coils are wound on the same armature, so that nearly continuous currents may be obtained; and the substance of the armature is iron, so that the induction through it is increased greatly owing to the magnetic permeability of the iron.

An elementary and simplified theory of the dynamo enables us to obtain a quantitative expression, approximately accurate, for the average value of the current. Let $N$ denote the magnetic flux, that is the total number of lines of magnetic induction which traverse the coil when normal to the lines of force. Then $-N$ must be the number through it if turned through 180°, when it lies again normal to the field with its faces reversed. The total change of flux is thus $2N$ during one half-revolution. If $t$ be the time occupied in this half-revolution, the average rate of change of flux is $2N/t$, and this measures the average induced electromotive force. If we neglect the change of sign of the current, or if we commutate the current so that it always passes in the same direction in the external circuit, we may calculate its average value by dividing the average electromotive force by $R$ the total resistance of the circuit, including both that in the coil and that without. If $x$ be the number of revolutions of the coil per second, $t$, the time of one half-revolution, is $1/2x$, and thus the average current $\bar{c}$ is given by

$$\bar{c} = \frac{4Nx}{R}.$$

If the total area of the coil, including the area of each of its windings, be $A$ ; and if $B$ be the magnetic induction in the field, assumed uniform, then

$$\bar{c} = \frac{4ABx}{R}.$$

In considering the output of power, it is necessary to calculate the value of the square root of the average value of $c^2$, instead of the average value of $c$, since the power of a current is $Ec$ or $c^2R$.

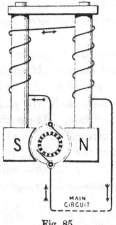

Fig. 85.

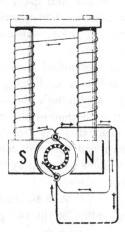

Fig. 86.

In alternating current machines, the field-magnets must be excited separately. In direct current dynamos, they may be excited by the current of the machine, by connecting the coils of the field-magnets either in series (Fig. 85) or as a shunt (Fig. 86) with the external circuit. In the shunt-wound machine, an increase of external resistance throws more current through the magnet coils, and thus increases the induced electromotive force. In series-wound machines, by increasing the external resistance we diminish the current everywhere, and thus decrease the induced electromotive force. By a proper combination of shunt and series coils on the field magnets, it is possible to secure a constant electromotive force, whatever be the resistance of the external circuit.

The magnetic flux through the armature produced by the current $c$ in the field magnet coils is a function of that current, and may be written as $\phi(c)$. It is equal to the electromotive force for one turn of the armature per second. For a speed of $n$ turns, the electromotive force will be $n\phi(c)$. The product $c\phi(c)$ represents the positive or negative work of the electromagnetic forces for one turn of the armature at any speed.

All the properties of the machine are known if the value of $\phi(c)$ is known. It is best represented by a curve drawn between $c$ as abscissa and $\phi(c)$ as ordinate. Such a curve is called a characteristic. Characteristic curves may be drawn for any given dynamo by measurements of current and electromotive force at known speeds.

The mechanical force acting on a current in a magnetic field is used to obtain mechanical power in electric motors. If a current be passed through the armature of a dynamo in the reverse direction to that of the current the machine itself will yield, it is evident that the armature will tend to rotate. At starting, a large current will pass, but, as the speed increases, a back electromotive force is set up by the usual action of the machine, and the current taken by the motor diminishes. Any dynamo will act as a motor, though the most convenient mechanical arrangements usually differ for the two purposes.

**59.** The power conveyed by an electric circuit, that is, the work done per second, is the product of the electromotive force and the current, or $Ec$. Thus, by increasing $E$, it is possible to convey a large amount of power by the use of a small current. The conductor required to carry the current with a certain percentage of heat-loss must possess an area of cross-section proportional to the current, and copper is very expensive. Hence it is economical to generate the current by a high tension dynamo, and convert the power into

*Transformers and induction coils.*

currents of a safe voltage near the consumer by means of an apparatus called a transformer.

In this instrument, two coils, one of few turns of thick wire, and one of many turns of thin wire, are wound on a core of iron wires or strips, which may either be straight or in the form of a continuous ring. An alternating current of high voltage is passed round the coil of many turns. The magnetism of the core is continually reversed by the alternating current, and the lines of induction through the other coil are reversed continually also. An induced electromotive force, proportional to the number of turns of wire, is obtained, and, as the resistance is low, a large current results. It is obvious that the power of the secondary current cannot exceed that of the primary, and we have as the limit of efficiency

$$E_1 c_1 = E_2 c_2.$$

If the circuit of the secondary coil be not completed, no current flows round it, and no power is taken from the transformer. We should suspect that in these conditions less power is taken from the primary circuit. This is found to be the case, and the reason is clear in the light of what was said on page 157. The presence of a closed secondary circuit in the neighbourhood decreases the effects of self-induction in the primary, and enables more current to pass through it. When the secondary circuit is broken, the primary acts as a choking coil (p. 161), and very little current flows, owing to the reverse electromotive force of self-induction.

The induction coil, invented by Ruhmkorff, may be classed as a transformer (Fig. 87). It enables us to obtain an intermittent current of high electromotive force from the direct current given by a few voltaic cells. Two coils are wound on an iron core made of a bundle of soft iron wires to reduce the loss by eddy currents, which would circulate extensively in a solid core. The inner coil is made of a few turns of thick wire, and the outer coil of an immense number of turns of very fine wire wound in sections, most carefully insulated, and so arranged that turns with high differences of potential are separated widely from each other. The primary current, which passes round the inner coil, is interrupted at frequent intervals of time. As a simple means of effecting this, the arrangement used in electric bells may be adopted. The primary current is made to pass from a platinum stud fixed to a spring into a platinum point in contact with it. To the end of the spring is attached a piece of soft iron, which lies near one end of the iron core of the coil. When the current passes, the core is magnetized, and attracts the piece of soft iron. The spring is pulled away from the point, and the circuit is broken. The spring then flies back, making contact with the point, and the current is established once more.

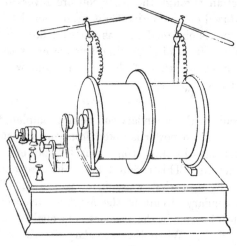

Fig. 87.

An electrolytic interruptor was invented by Wehneldt, who found that, if the primary current were passed through a solution of dilute sulphuric acid, the electrodes being a platinum point and a platinum plate, rapid interruptions occurred spontaneously. The action of this apparatus is not fully understood. It may depend on sudden evolutions of gas at the electrodes.

By some such automatic arrangement, the primary current is made and broken rapidly, and a series of alternately directed electromotive forces are set up in the secondary coil. Since the tubes of magnetic induction produced by the primary current must thread the many turns of the secondary circuit, the induced electromotive force is very large. Such a coil, then, would furnish an alternating current in the secondary circuit. By the use of a condenser, however, we may increase the electromotive force which arises in the secondary circuit when the primary current is broken. The condenser is made of intervening sheets of tin-foil and paraffined paper, the alternate sheets of foil being connected together. Its opposite systems of plates are connected one with each side of the contact-breaker: in the usual form, one with the platinum point and one with the spring. When the point is in contact with the spring, the condenser is discharged and the core is magnetized. When the point separates, the condenser is charged through the primary coil, the process taking place by a series of electric oscillations which rapidly demagnetize the iron core. By this means the flux of induction through the coil is destroyed much more sharply than it would be in the absence of a condenser, and the electromotive force on breaking the current is increased greatly.

The use of induction coils has extended considerably of late, owing to their employment in hospitals for the production of Röntgen rays, and their application to many of the more recent branches of physical research.

## REFERENCES.

*Elements of the Mathematical Theory of Electricity and Magnetism*; by Sir J. J. Thomson; Chapter XI.
*The Theory of Alternating Currents*; by Alexander Russell.
*Dynamo-Electric Machinery*; by S. P. Thompson.
Articles "Dynamo," etc. in the *Encyclopædia Britannica*.
Articles in the *Dictionary of Applied Physics*.

# CHAPTER VIII

## ELECTRICAL UNITS

The units of physical measurement. Electrostatic and electromagnetic units. The determination of a current in absolute measure. The determination of a resistance in absolute measure. The determination of an electromotive force in absolute measure. Other practical units.

**60.** PHYSICAL science is based on the perceptions of our senses, and the quantities with which we deal in studying that branch of knowledge are based, more or less directly, on such sense-perceptions. In measuring a physical quantity, two factors are necessary. We must choose and define some unit of the quantity, and we must determine how many times that unit is comprised in the quantity we wish to measure.

The units of physical measurement.

The ideas of length and time may be regarded as primary— length as the simplest form of our conception of space, time as a recognition of sequence in our states of consciousness. It is usual to take mass as a third fundamental quantity. Whether mass is to be considered as truly a primary conception is still a matter of controversy. In the view of the present writer, the third primary conception is that of force, which we derive from our muscular sense when pushing or pulling. Equal masses are then defined as masses in which equal forces produce equal accelerations, and unequal masses are in the inverse ratio of the accelerations so produced.

The conception of mass, whether it is primary or derived from that of force, is soon found necessary in the mental picture which the study of dynamics enables us to make of the world around us. The great convenience of the conception of mass arises from the fact that, in any system, the quantity so named is found to remain constant throughout a series of dynamical changes. On account of this principle of the conservation of mass, it is well to take mass as the third fundamental physical quantity, and base our system of units on those of length, time and mass.

These fundamental units must be selected arbitrarily. In civilised countries, the unit of length is taken as the length between two marks on a certain standard metallic bar. In England there is a standard yard, and in France a standard metre. In point of fact, both these units are selected arbitrarily for their convenience, though the original idea of the metre was derived from a connexion with the supposed dimensions of the earth. For scientific purposes, the hundredth part of the metre, or centimetre, is taken as the unit of length.

Like the unit of length, the unit of time is arbitrary, and for the convenience of daily life the obvious unit to select is the mean solar day, while the sequence of the seasons suggests another equally arbitrary unit, the year. The exact relation between these two units can only be determined by careful astronomical observation. Again, for laboratory use, a smaller unit is convenient, and the second, a certain fraction of the astronomical unit, is chosen.

The unit of mass, whether pound or kilogramme, likewise, is fixed arbitrarily, and defined as a mass equal to that of a standard kept at some Government Office, though the original kilogramme was made as nearly equal as was then possible to the mass of one cubic decimetre of water. Once more, a smaller unit is convenient in the laboratory, and the thousandth part of the kilogramme, or gramme, has been chosen. The system of units based on the centimetre, the gramme, and the second, is in almost universal use for scientific purposes, and may be referred to as the C.G.S. system.

All other dynamical units may be derived from these three fundamental units of length, time, and mass. Thus, the unit of velocity is the velocity of one centimetre per second, and that of acceleration means unit increase of velocity per second, or the

acceleration of one centimetre per second per second. The
definition of mass (p. 176) gives us the relation

$$\text{mass} = \text{force/acceleration,}$$

and, therefore, taking mass as our fundamental unit for the sake
of using a constant quantity, we see that the unit of force, in the
c.g.s. system called the dyne, is that force which produces unit
acceleration when acting on unit mass. We may express this
result in the form

$$[\text{Force}] = M\,[\alpha] = MLT^{-2},$$

a relation that gives us the physical dimensions of force in terms
of the three fundamental units. Also we see that the dimensions
of work or energy are given by the relation

$$[\text{Work}] = ML^2T^{-2}.$$

**61.** When we pass to the consideration of electrical quantities,
Electrostatic we find that, in the present state of science, we cannot
and electro-
magnetic express them completely in terms of dynamical units.
units. It is possible that some day this may be done, but,
when the necessary relation has been discovered, it will not follow
that it is better to express electrical quantities in terms of those
of dynamics. It may turn out to be quite as correct philosophically
to express dynamical units in terms of electricity, though the
absence of any special electrical sense may make such a proceeding
less convenient and satisfying to the human mind. It should be
pointed out, however, that, to the electrical fish or torpedo, electric
intensity may be a direct sense-perception, and quite as real as
mechanical force.

On the electrostatic system of units, we ought strictly to define
unit charge of electricity by means of the experimental relation
given on page 31,

$$F = \frac{e_1 e_2}{k r^2},$$

where $F$ is the mechanical force between two quantities $e_1$ and $e_2$
of electricity, which as point-charges are placed in a medium of
dielectric constant $k$ at a distance $r$ from each other.

Hence, the dimensions of quantity of electricity on the electro-static system are

$$[e] = M^{\frac{1}{2}} L^{\frac{3}{2}} T^{-1} k^{\frac{1}{2}},$$

and involve the unknown dimensions of the specific inductive capacity or dielectric constant of the medium, though the conventional unit is defined on the supposition that this unknown quantity has no dimensions.

In a similar manner, we may deduce the dimensions of the other electrostatic units, and tabulate them as follows:

| | | | | | |
|---|---|---|---|---|---|
| Dielectric constant ... | ... | ... | ... | $k$ | $k$ |
| Quantity of electricity | ... | ... | ... | $e$ | $M^{\frac{1}{2}} L^{\frac{3}{2}} T^{-1} k^{\frac{1}{2}}$ |
| Electric intensity ... | ... | ... | ... | $f$ | $M^{\frac{1}{2}} L^{-\frac{1}{2}} T^{-1} k^{-\frac{1}{2}}$ |
| Potential difference ... | ... | ... | ... | $V$ | $M^{\frac{1}{2}} L^{\frac{1}{2}} T^{-1} k^{-\frac{1}{2}}$ |
| Capacity ... | ... | ... | ... | $C$ | $Lk$ |
| Electrostatic polarization ... | ... | ... | $P$ | $M^{\frac{1}{2}} L^{-\frac{1}{2}} T^{-1} k^{\frac{1}{2}}$ |
| Current ($=e$/time) ... | ... | ... | ... | $c$ | $M^{\frac{1}{2}} L^{\frac{3}{2}} T^{-2} k^{\frac{1}{2}}$ |
| Strength of magnetic pole ($=$work/current) | $m$ | $M^{\frac{1}{2}} L^{\frac{1}{2}} k^{-\frac{1}{2}}$ |
| Magnetic force ... | ... | ... | ... | $H$ | $M^{\frac{1}{2}} L^{\frac{1}{2}} T^{-2} k^{\frac{1}{2}}$ |
| Magnetic permeability ... | ... | ... | $\mu$ | $L^{-2} T^2 k^{-1}$ |

The electromagnetic system of units should, in like manner, be based on the expression for the force between two magnetic poles placed in a medium of magnetic permeability $\mu$, at a distance $r$ from each other:

$$F = \frac{m_1 m_2}{\mu r^2},$$

though in our conventional definitions we agree to ignore the unknown dimensions of the magnetic permeability of the medium.

Unit magnetic force or field is such that it exerts unit force on a magnetic pole of unit strength. In the C.G.S. system, unit magnetic field is called the Gauss, and unit magnetic flux the Maxwell.

Unit current is defined as that current which involves the performance of work equal to $4\pi$ ergs when unit magnetic pole is carried round it (p. 103). The derivation of the other units which are tabulated below will be clear.

| | | |
|---|---|---|
| Magnetic permeability | $\mu$ | $\mu$ |
| Strength of magnetic pole | $m$ | $M^{\frac{1}{2}} L^{\frac{3}{2}} T^{-1} \mu^{\frac{1}{2}}$ |
| Magnetic force | $H$ | $M^{\frac{1}{2}} L^{-\frac{1}{2}} T^{-1} \mu^{-\frac{1}{2}}$ |
| Magnetic potential | $V'$ | $M^{\frac{1}{2}} L^{\frac{1}{2}} T^{-1} \mu^{-\frac{1}{2}}$ |
| Magnetic induction | $B$ | $M^{\frac{1}{2}} L^{-\frac{1}{2}} T^{-1} \mu^{\frac{1}{2}}$ |
| Current ($=$work/pole) | $c$ | $M^{\frac{1}{2}} L^{\frac{1}{2}} T^{-1} \mu^{-\frac{1}{2}}$ |
| Quantity of electricity <br> ($=$current $\times$ time) | $e$ | $M^{\frac{1}{2}} L^{\frac{1}{2}} \mu^{-\frac{1}{2}}$ |
| Electric potential-difference <br> or electromotive force | $V$ or $E$ | $M^{\frac{1}{2}} L^{\frac{1}{2}} T^{-2} \mu^{\frac{1}{2}}$ |
| Coefficient of self or mutual <br> induction ($=$work/(current)$^2$) | $L$ or $M$ | $L\mu$ |
| Electric capacity | $C$ | $L^{-1} T^2 \mu^{-1}$ |
| Dielectric constant | $k$ | $L^{-2} T^2 \mu^{-1}$ |

Thus, in each system of units, the unknown dimensions of a property of the dielectric medium are involved—its specific inductive capacity on the one side, and its magnetic permeability on the other. The accepted definitions of the various units ignore these unknown quantities, and treat $k$ and $\mu$ as numbers having no dimensions—as mere ratios, giving the value of the quantity in terms of that of air. The unit of electric charge, for instance, is defined as that quantity of electricity which repels an equal similar quantity with the force of one dyne when each of the two quantities is placed as a point-charge in air, at a distance of one centimetre from the other.

We now see that, by ignoring the unknown dimensions of $k$ and $\mu$, it is possible to arrive at a definition of any electrical or magnetic quantity by two distinct methods. Two distinct units result—units different in dimensions and different in magnitude, the difference arising from the assumption that $k$ and $\mu$ have no dimensions. Let us take, for example, the unit of current. On the electrostatic system, the idea of current is derived from that of electric charge—unit current flows through a conductor when unit charge passes through it at a uniform rate in unit time. On the electromagnetic system, on the other hand, a current is detected, defined and measured by means of its magnetic effects; the conception of a quantity of electricity is now derived from that of current. The electromagnetic unit of current is the current which involves $4\pi$ ergs of work when unit pole is carried round it, or the current which produces a magnetic force equal to $2\pi/r$ at the centre of a circular path of radius $r$. By either of these definitions, we obtain a unit of current different from that derived from the electrostatic system. The electrostatic dimensions of current, as usually defined, are those of quantity of electricity divided by time, or $M^{\frac{1}{2}}L^{\frac{3}{2}}T^{-2}$, while the electromagnetic dimensions are $M^{\frac{1}{2}}L^{\frac{1}{2}}T^{-1}$. Thus, the units of current as ordinarily defined possess different dimensions on the two systems, and the ratio of the dimensions of the electrostatic to those of the electromagnetic unit is

$$\frac{M^{\frac{1}{2}}L^{\frac{3}{2}}T^{-2}}{M^{\frac{1}{2}}L^{\frac{1}{2}}T^{-1}} = LT^{-1}.$$

Now, $LT^{-1}$ is the dimensions of a velocity. Hence, the ratio of the electrostatic to the electromagnetic unit of current involves the dimensions of a velocity.

Again, on the electrostatic system, the dimensions of capacity, namely, quantity of electricity divided by potential-difference, are simply those of length, $L$, while, on the electromagnetic system, they are $L^{-1}T^2$. Here the ratio of the units, electrostatic to electromagnetic, involves $L^2T^{-2}$, that is, the square of a velocity.

These discrepancies between the units of one and the same quantity depend, as we have seen, on the assumption that the specific inductive capacity and the magnetic permeability of the electromagnetic medium are mere ratios—quantities of no physical dimensions, expressing the property of the medium simply as compared with the same property of air. If, instead of thus ignoring the dimensions of $k$ and $\mu$, we keep these symbols, unknown as they are, in our definitions, we may assume that the same quantity, however defined, will possess the same dimensions. Taking for instance the two definitions of unit current, we have

$$M^{\frac{1}{2}}L^{\frac{3}{2}}T^{-2}[k]^{\frac{1}{2}} = M^{\frac{1}{2}}L^{\frac{1}{2}}T^{-1}[\mu]^{-\frac{1}{2}}.$$

Thus, $$[k]^{\frac{1}{2}}[\mu]^{\frac{1}{2}} = L^{-1}T,$$

and a similar result is reached whatever pair of units we consider.

Hence, although the individual dimensions of the two quantities,

the dielectric constant and the magnetic permeability, remain
unknown, it follows that the dimensions of their product are
determined, and are given by the relation

$$[k\mu] = \frac{1}{v^2}.$$

To this important result we shall return later; our immediate
concern is with the methods of determining experimentally the
values of different electrical quantities in the absolute electrostatic
or electromagnetic units defined on the assumption that $k$ and $\mu$
are quantities having no dimensions.

**62.** The definition of unit current as equivalent to a magnetic
shell of unit strength has been shown in § 41,
Chapter V, to lead to the same results as a more
practical definition, which makes the unit current
that current which, when owing round a circle of
radius $r$, produces a magnetic force of $2\pi/r$ at the centre. From
this definition we deduced the theory of the tangent galvanometer,
and showed that, with proper precautions, that instrument gave a
fairly accurate means of determining the electromagnetic value of
a current. Unfortunately, this method involves a knowledge of the
horizontal component of the earth's magnetic force; and, not only
is the accurate determination of this quantity a long and laborious
undertaking, but the continual variations of its value make it
necessary to record its changes throughout the time of the
experiments. This objection applies, it is true, to some of the
further methods now to be discussed, but others, less affected by
the difficulty, have been devised.

The determination of
a current in
absolute
measure.

One such method depends on the measurement by weighing
of the force between circuits when currents flow round them.
Practical instruments have been devised by Lord Kelvin and others
for the measurement of currents by means of such forces. In
these instruments, two small moveable coils are attached one to
each end of the beam of a balance. Each of the moveable coils
lies between and parallel to two larger fixed coils of approximately equal size, round which the currents pass in such a
direction that the force on one moveable coil acts upwards and

the force on the other coil acts downwards (Fig. 88). A
pointer at the end of
the beam shows when
it is horizontal, and a
sliding weight can be
adjusted to bring the
beam back to this
position when it is
deflected. The couple
tending to deflect the

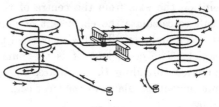

Fig. 88.

beam varies as the square of the current, while that tending to
restore the beam to its position of equilibrium varies as the displacement of the sliding weight. Hence the current is proportional to the
square root of the distance between the weight and the fulcrum.

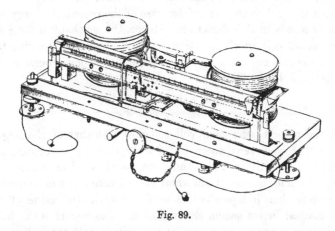

Fig. 89.

The beam is graduated according to this square root scale, and,
if we know the absolute value of the graduations, the instrument
may be used as a standard of reference. It is called an ampere-
balance.

For our present purpose of investigating the absolute electromagnetic value of some current which can afterwards be recovered,
we must calculate the couple on the moveable coils from the
dimensions and arrangement of the apparatus. We have seen, § 41,
that the magnetic force on the axis of a large circular current is

equal to $2\pi n_1 c r_1^2/(r_1^2 + x^2)^{\frac{3}{2}}$, where $c$ is the current, $n_1$ the number of turns of wire, $r_1$ the radius of the coil, and $x$ the distance of the point on the axis from the centre of the coil. We may use this result to calculate the coefficient of mutual induction between a large coil and a small coil with parallel planes, the two coils being coaxial. The coefficient $M$ is the number of tubes of magnetic induction threading the total area $n_2\pi r_2^2$ of the small coil when unit current circulates in the large coil. It will therefore have the value

$$n_2\pi r_2^2 \cdot 2\pi n_1 r_1^2/(r_1^2 + x^2)^{\frac{3}{2}}$$

or

$$2\pi^2 n_1 n_2 r_1^2 r_2^2/(r_1^2 + x^2)^{\frac{3}{2}}.$$

Now we have seen that the mutual potential energy of two circuits, when unit current passes through both, is measured by their coefficient of mutual induction. The force on a body in any direction is equal to the rate of decrease of the potential energy as the body moves in that direction. Hence, the mechanical force $F$ on the small coil in the position we have described, tending to move it towards the large coil when a current $c$ flows round both coils, is $- c^2 dM/dx$. Thus

$$F = 6\pi^2 n_1 n_2 r_1^2 r_2^2 \frac{x}{(r_1^2 + x^2)^{\frac{5}{2}}} c^2.$$

If the plane of the small coil coincide with that of the large one, so that $x$ is zero, there is no force tending to change $x$, while at a great distance also the force vanishes. Hence there is some position for the small coil where the attraction is a maximum; the force is then independent of small errors in the value of $x$. Mathematical investigation shows that this position is such that the distance between the planes of the coils is half the radius of the larger coil, and that the force between the coils is given by the expression

$$F = \frac{96\pi^2 n_1 n_2 r_2^2}{5^{\frac{5}{2}} r_1^2}.$$

Thus the result depends on the ratio of the effective radii of the large and small coils, not on the absolute values of those radii. Now the ratio of the radii may be found by measuring electrically the ratio of the galvanometer-constants of the two coils, a quantity which can be obtained much more accurately than the effective radius of a number of turns of wire wound in a coil.

The approximate theory of the subject given above assumes that each coil may be treated as a circle, the dimensions of the channel for the wires being supposed negligible compared with the radius of the circle. This, of course, is very far from being the case, and the exact theory of the apparatus is somewhat complicated. Nevertheless, the investigation we have given serves to illustrate the principles involved.

The method was used by Lord Rayleigh and Mrs Henry Sidgwick in 1884 for the determination of a current in absolute measure. The small coil was suspended from one arm of a balance, the current being taken in by means of flexible leads of fine copper wire. The large coil, as in the ampere-balance described above, was doubled; one large coil was placed on each side of the small coil, and the current was sent opposite ways round the two large coils. The force was thus doubled, and made more independent of the exact position of the small coil.

When a resistance has been determined in absolute measure, it is possible to construct a standard resistance coil and keep it for future reference. Accurate copies of such coils are easily made at a comparatively small cost. Standard ampere balances are now set up in the various National Laboratories, but they are complicated and expensive instruments. It is well, therefore, at the time of an absolute determination, to measure the same current in some practical way, which may afterwards be repeated with other currents, in order to obtain an indirect measurement of those currents in absolute units.

One property of a current, which has often been used for this purpose, is its power of depositing a metal from the solution of one of its salts. This process will be studied in the chapter on electrolysis. Here it is enough to say that the amount of silver deposited from a solution of silver nitrate by a given current flowing for a given time, that is, by the passage of a given quantity of electricity, is a quantity which, within the limits of experimental accuracy, seems to be quite constant. The mass of silver deposited, then, may be used as a measure, not directly

indeed of the current passing at any particular instant, but of
the total electric transfer round the circuit during the time of
the observation.  If, however, the current be maintained constant
throughout, its value at any instant will be proportional to the
mass of silver divided by the time.

The experiments of Lord Rayleigh and Mrs Sidgwick went to
show that one electromagnetic unit of electricity, when passed
through a solution of silver nitrate from a plate of silver to the
surface of a platinum bowl, deposited 0·0111795 gramme of silver.
Many later experiments have been made, with the result that the
mean of the best observations indicates that the true figure is
probably between 0·01118 and 0·01119.  The ampere, or practical
unit of current, is one-tenth of the electromagnetic unit, and has
been defined for practical purposes by the International Congress
which met at Chicago in 1893 as that current, which, flowing
through a solution of silver nitrate, deposits 0·001118 gramme of
silver per second.  This is now the legal definition of the ampere,
and it is possible that, even should future research show that the
electromagnetic unit is appreciably different from the value taken
at present, no change would be made in the practical definition of
the ampere.  The supposed dimensions of the earth, assumed for
the original construction of the standard metre by the reforming
zeal of the first French Republic, have been shown by later
determinations to be erroneous; but the distance between two
marks on the French standard bar then made continues to be
the legal metre.

Another method, for the determination of currents in absolute
measure, depends on the use of an instrument known as Weber's
electrodynamometer.  Two large circular coils are mounted with
their planes in the magnetic meridian, the coils being coaxial.
They may be looked upon as part of a double-coil tangent galvanometer.  Instead of the needle, however, a smaller coil is hung by a
bifilar suspension at the middle of the line joining the centres of
the two coils.  The plane of the small coil, when in its position
of equilibrium, lies at right angles to the planes of the two
large coils.  A current is passed in series through the two large
coils, and, by means of the bifilar suspension, through the

small suspended coil, which then tends to set with its plane parallel to those of the large fixed coils, as may be seen by considering the direction of the axis of the equivalent magnetic shell.

The couple of restitution due to the bifilar suspension is approximately proportional to the sine of the angle of deflection, and may be determined by experiment: for a deflection $\theta$ let us call it $b \sin \theta$. The couple due to the current is

$$\frac{2\pi n_1}{r_1} cM \cos \theta,$$

where $M$ is the magnetic moment of the suspended coil. The strength of the current is measured by the strength of the equivalent magnetic shell, and the strength of a shell is the magnetic moment per unit area. Thus, the effective area of the suspended coil being $\pi r_2^2 n_2$, its magnetic moment is $\pi r_2^2 n_2 c$. We therefore have

$$\frac{2\pi n_1}{r_1} c\pi r_2^2 n_2 c \cos \theta = b \sin \theta,$$

or

$$c^2 = \frac{b r_1}{2\pi^2 r_2^2 n_1 n_2} \tan \theta.$$

When a current flows in the suspended coil, the coil tends to set in the earth's field, and the couple due to this tendency must be added to that of the bifilar suspension. Corrections are also necessary for the finite size of the windings, and other differences between the practical apparatus and the simplified form of it contemplated by the elementary theory given above.

It should be noted that the deflection depends on $c^2$, so that the electrodynamometer may be employed for alternating currents.

**63.** The three quantities, current, electromotive force, and resistance, are connected together by the well-known relation described by Ohm's law. Hence the electri-

The determination of a resistance in absolute measure.

cal quantities involved in the passage of a steady current are all determined in absolute electromagnetic measure if the values of two of these three quantities are known. We have described methods of determining the strength of a current, and it remains to consider the second type of experiment, whereby the solution of the problem is completed. The

experiments give values both of the electromotive force and also
of the resistance, since, when the current is known, there is a
necessary relation between them. But, since permanent resistance
coils are easy to construct, it is usual to regard the experiments as
an investigation of a resistance in absolute measure, or, as the
phrase goes, the determination of the ohm.

The first determination of the resistance of a wire in absolute
electromagnetic measure was made by Kirchhoff, who calculated
from their dimensions the coefficient of mutual induction of two
coils when in a certain relative position, and observed the throw
of a galvanometer when one coil was moved quickly into a position
where the mutual induction was zero.

Joule's calorimetric experiments on the heat developed by a
current (see page 135) give a value for the work done by a current
when passing through a certain resistance. The current may
be measured in electromagnetic units by means of a tangent
galvanometer or otherwise, and the observed thermal effect
then gives the absolute value of the electromotive force and
the resistance between the ends of the coil immersed in the
calorimeter.

The first determination, on which was based an extensive
manufacture of resistance coils, was undertaken at the instance
of the British Association, and was carried out by Balfour Stewart,
Fleeming Jenkin, and Clerk Maxwell in 1863.

In order to understand the principles involved, let us consider
the case of a coil of wire, spinning round a vertical axis in the
earth's magnetic field of force. The plane of the coil being
vertical, the vertical magnetic component will not be involved, and
may be neglected; the horizontal component alone need be taken
into account. As we saw in § 58, when considering the theory of
the dynamo, a periodic electromotive force is set up in the coil,
the direction being reversed twice in each complete revolution,
when the plane of the coil lies at right angles to the lines of
induction. An alternating current will circulate in the coil; it
will have a periodic time equal to that of the electromotive force,
and, owing to the small self-induction of the coil, will not lag
appreciably behind the electromotive force in phase.

An alternating current of short period will not affect an

ordinary galvanometer. By leading off the current into an external circuit through a rectifying commutator, as described on page 170, its average value might be determined by a tangent galvanometer. But such a measurement could give only a roughly accurate result, and another method was adopted.

If a compass-needle be suspended at the centre of the coil itself, the position of the coil relative to that of the needle will be reversed periodically in time with the reversals of the current. Thus the couple on the needle due to the current will act always in the same direction, and the needle will suffer a permanent deflection, which is constant if the coil rotate at a constant rate, of which the period is short compared with the time of vibration of the needle itself.

The average couple on the needle due to the varying current in the coil may be calculated, and thus its observed deflection may be used to determine the average current in electromagnetic units. It will be noticed that the horizontal component of the earth's magnetic field comes into each side of the equation of equilibrium of the needle; it determines the induced electromotive force in the coil, and the couple tending to deflect the needle; but also it determines the couple tending to return the needle to its original position of equilibrium. Therefore, to a first approximation, the indications of the apparatus are independent of the value of the earth's field. A correction has to be applied, however, for the magnetic field of the needle itself, which of course produces an electromotive force in the spinning coil.

The electromotive force in the coil being thus calculated, and the current being deduced in electromagnetic units from the observed deflection of the needle, the resistance of the coil can be calculated in absolute units, and, by comparison with it, resistance coils of any required value may be constructed. The ohm coils based on the results of these experiments—the B.A. units, as they are often called—for long were the usual standards of resistance. Of recent years, however, more accurate determinations of the ohm, both by the use of spinning coils and by other methods now to be described, have led to the construction of new standards. The old B.A. unit is 0·9863 of the new or "international ohm."

Perhaps the method which is capable of determining a resistance in electromagnetic units with the greatest precision is that due to Lorenz, and used by him, by Lord Rayleigh and Mrs Sidgwick, and by others, for the determination of the ohm.

We have seen (page 153) that, if part of a circuit be moveable, it will tend to travel under the influence of a magnetic field so that the total induction through the circuit is increased. Let us imagine that a metallic disc is mounted to rotate on its axis so that its plane is normal to a magnetic field of known intensity, and that two brushes touch the disc, one at its axis and one at its circumference. If a current be passed through the disc between these brushes, the magnetic force will so act on each element of current that it tends to move round the axis of the disc. Thus, on the disc as a whole, there act forces with a moment about the axis tending to make the disc rotate in its own plane.

Hence it follows that, in the converse case, when no external current is passed through the disc, but the disc is made to rotate by external means in the magnetic field, an electromotive force will be set up in the circuit, consisting of the disc, the brushes, and an external wire joining the brushes.

We may calculate the electromagnetic value of this electromotive force, by considering the work done by the battery which we imagined to maintain the current in the first case; or, more simply, by reckoning the number of magnetic lines of induction cut by the lines of current-flow in the disc when it is made to rotate as supposed in the second case. The current passes from the axis of the disc to its circumference, and any line of current-flow, whether straight or curved, will, at each revolution, cut all the tubes of induction which pass through the disc. Thus the total change of induction per second through the circuit, which measures the induced electromotive force $E$, is given by the relation

$$E = nN,$$

where $n$ denotes the number of revolutions made by the disc per second, and $N$ is the total flux of induction through the disc.

Lord Rayleigh placed the disc between two circular coils through which passed a constant current, the axes of the coils and of the disc being coincident. If, for the moment, we neglect

the magnetic field of the earth, which is small compared with that
due to the two coils, we see that

$$N = cM,$$

where $M$ is the coefficient of mutual induction between the coils
and the edge of the disc, and $c$ the current passing through them.
The electromotive force, due to the rotation of the disc, is now
given by

$$E = ncM.$$

The arrangement of apparatus is shown in Figure 90. A
galvanometer (1) is placed in the circuit connecting the centre of

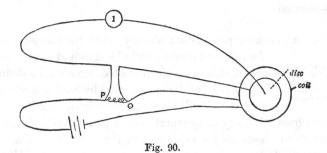

Fig. 90.

the disc and the rim, and the electromotive force in this circuit is
balanced against that in a certain part $PQ$ of the main circuit of
the two coils, the resistance between $P$ and $Q$ being adjusted till
no current passes through the galvanometer. If $R$ be the resistance between $P$ and $Q$, we then have

$$E = Rc = ncM$$

or     $$R = nM.$$

The value of $M$, the coefficient of mutual induction between
the two coils and the edge of the disc, can be calculated from their
dimensions and relative positions, and hence we obtain the resistance
$R$ in electromagnetic measure when the speed of rotation is known.

In a method used by Sir Richard Glazebrook for the determination of the ohm, the mutual induction between two coils is calculated
from their dimensions and relative positions, and then determined

experimentally by observing the throw of a ballistic galvanometer. The galvanometer is standardised by passing through it a steady current furnished by the electromotive force between two points in the primary circuit.

On the mean result of the best determinations, the International Congress of 1893 re-defined the ohm. One ohm is $10^9$ electromagnetic C.G.S. units, and, for practical purposes, may be taken as the resistance at the temperature of melting ice, of a column of mercury of uniform cross-section, which, at the same temperature, has a length of 106·3 centimetres, and a mass of 14·4521 grammes. Such a column is, as nearly as possible, one square millimetre in cross-section

**64.** As we have pointed out already, when the absolute values of a standard current and of a standard resistance are known, that of an electromotive force can be deduced. One method of doing this, used by Sir Richard Glazebrook and Mr Skinner, may briefly be described. The current from a battery of accumulators is passed through a silver voltameter to measure its value in absolute units by reference to the standard determinations. The current also passes through a resistance known in terms of the standard ohm. This resistance is made of a thin but broad strip of platinoid alloy which is not appreciably heated by a current of about an ampere. If $R$ be the resistance between the terminals of the strip and $c$ the current through it, the electromotive force between its terminals is $Rc$, and this electromotive force may be balanced directly against that of some adjustable electromotive force, as in the potentiometer. The adjustable electromotive force is obtained by passing the current from two Leclanché cells through two resistance boxes in series. The sum of the resistances in the two boxes is maintained constant, so that, if resistances are taken out of one box, equal resistances are put into the other. The current is thus kept constant, while the electromotive force between the terminals of one box can be adjusted at will. This electromotive force is balanced first against the electromotive force between the ends of the strip, which is known in C.G.S. units, and then against that of

*The determination of an electromotive force in absolute measure.*

some constant cell, such as that of Latimer Clark, by introducing
them in opposite directions into the same circuit of a high resistance reflecting galvanometer, and adjusting the resistances in the
boxes till there is no deflection. The electromotive force of the
Clark's cell is then known in terms of the electromotive force
between the ends of the strip, and thus its value is obtained in
absolute measure. The cell may afterwards be used as a standard,
and the electromotive forces of other cells determined by comparison with it. The comparison may be effected by means of the
potentiometer in the usual manner.

The volt is $10^8$ electromagnetic units, and the Congress of 1893
defined the International Volt as the electromotive force, which,
steadily applied to a conductor with a resistance of one international ohm, will produce a current of one international ampere,
and is represented sufficiently well for practical use by 1000/1434
of the electromotive force at 15° C. between the poles of a Clark's
cell prepared according to a standard specification. Since the date
of the Congress several new investigations have been made, and it
seems possible that the true electromotive force of a Clark's cell is
lower by one or two parts in a thousand, and more nearly equal to
1·432 or 1·433 volts.

**65.** When the absolute measures of the practical units of
Other practical current, resistance, and electromotive force are once
units. determined, the values of the practical units of the
other electrical quantities follow by definition, and may be determined experimentally by ordinary electrical methods without new
absolute measurements. The following definitions were adopted
in 1893.

*Quantity.* The International Coulomb is the quantity of electricity transferred by a current of one international ampere in one
second.

*Capacity.* The International Farad is the capacity of a condenser which is charged to a potential of one international volt by
one international coulomb of electricity.

*Work.* The Joule is equal to $10^7$ ergs (units of work in the
C.G.S. system), and is represented sufficiently well for practical use

by the energy expended in one second by an international ampere in an international ohm.

*Power.* The Watt is equal to $10^7$ units of power in the C.G.S. system, that is, $10^7$ ergs per second, and is represented sufficiently well for practical use by the work done at the rate of one Joule per second.

*Inductance.* The Henry is the coefficient of induction in a circuit when an electromotive force induced in this circuit is one international volt, while the inducing current varies at the rate of one ampere per second.

## REFERENCES.

*Encyclopædia Britannica*, Art. "Electric Units."
*Absolute Measurements in Electricity and Magnetism.* A. Gray.

# CHAPTER IX

## ELECTROMAGNETIC WAVES

The electromagnetic medium. The propagation of an electromagnetic disturbance. The ratio of the electrostatic and the electromagnetic units. The relation between dielectric constants and optical refractive indices. The experiments of Hertz. Wireless telegraphy. The origin and mode of propagation of waves of light. The energy of the electromagnetic field. The momentum in the electromagnetic field. Magnetic forces as due to the motion of electrostatic tubes of force.

**66.** FARADAY was the first to fix his attention on the dielectric medium as the essential seat of electrical processes. As we saw in Chapter III, it was to Faraday's "obstinate disbelief" in action at a distance that we owe most of his researches. Faraday had no skill in mathematical analysis, and his conceptions were much in advance of the general knowledge of his time. As was said by the great German physicist von Helmholtz in his Faraday Lecture for 1881, "Now that the mathematical interpretation of Faraday's conceptions regarding the nature of electric and magnetic forces has been given by Clerk Maxwell, we see how great a degree of exactness and precision was really hidden behind the words, which, to Faraday's contemporaries, appeared either vague or obscure."

Taking up Faraday's ideas, and developing them in new directions, Maxwell pointed out that, if the energy of an electromagnetic system resided in the dielectric medium, it was probable that the energy passed through that medium with a finite velocity. When, for instance, an alternating current in one circuit is inducing a secondary alternating current in another circuit at a distance, the energy must pass through the intervening space in some alternating form also. A wave motion of some kind is thus suggested, and a medium is required to carry the waves—to supply what the late Lord Salisbury once called "a nominative case to the verb 'to undulate.'"

In the chapter of his great book on Electricity and Magnetism, wherein Clerk Maxwell gave to the world his complete theory of electromagnetic waves, he writes:

"In several parts of this treatise an attempt has been made to explain electromagnetic phenomena by means of mechanical action transmitted from one body to another by means of a medium occupying the space between them. The undulatory theory of light also assumes the existence of a medium. We have now to show that the properties of the electromagnetic medium are identical with those of the luminiferous medium.

"To fill all space with a new medium whenever any new phenomenon is to be explained is by no means philosophical, but if the study of two different branches of science has independently suggested the idea of a medium, and if the properties which must be attributed to the medium in order to account for electromagnetic phenomena are of the same kind as those which we attribute to the luminiferous medium in order to account for the phenomena of light, the evidence for the physical existence of the medium will be considerably strengthened.

"But the properties of bodies are capable of quantitative measurement. We therefore obtain the numerical value of some property of the medium, such as the velocity with which a disturbance is propagated through it, which can be calculated from electromagnetic experiments, and also observed directly in the case of light. If it should be found that the velocity of propagation of electromagnetic disturbances is the same as the velocity of light, and this not only in air, but in other transparent media, we shall have strong reasons for believing that light is an electromagnetic phenomenon, and the combination of the optical with the electrical evidence will produce a conviction of the reality of the medium similar to that which we obtain, in the case of other kinds of matter, from the combined evidence of the senses."

As we shall see, this identity of velocity has been demonstrated by experiment, and, as Maxwell says, the evidence for the existence of a luminiferous and electromagnetic medium—an aether—is the same in kind, perhaps as strong in degree to those capable of judgment, as the evidence for the existence of other kinds of matter which we can touch and handle. The question whether this evidence is enough to demonstrate the real physical existence of entities corresponding to our conceptions of the aether or a table remains an inquiry for that branch of philosophy known as metaphysics. Experience demands the conceptions of aether and table to bring order into our mental picture of the universe, but natural science alone is unable to decide whether definite and ultimate realities exist, corresponding to our conceptions. The unity and simplicity which these conceptions introduce into our model of phenomena are valid metaphysical arguments in favour of their ultimate existence, but the arguments are metaphysical arguments, not scientific ones, in the limited sense of that word. Most men of science are realists, but their beliefs on such questions are metaphysical hypotheses or philosophic dogmas, and do not follow necessarily from the results of natural science.

**67.** Let us now consider a simple electric system by the properties of which we may investigate the propagation of an electromagnetic disturbance through an isotropic medium. Let $A$ and $B$ represent two vertical metallic plates very close together, forming a condenser charged with equal and opposite quantities of electricity uniformly distributed over each plate. The dielectric medium between the plates is in a state of strain, and straight horizontal Faraday's tubes of force may be supposed to run from the one plate to the other.

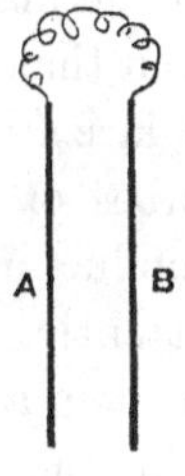

*The propagation of an electromagnetic disturbance.*

Fig. 91.

If the top of the plates be connected by means of a wire of high resistance, the plates are discharged. There is a flow of positive electricity up one plate, and of negative electricity up the other. Corresponding with this process, the electric tubes of force move vertically upwards towards the wire, in which they contract and disappear, as described on page 59.

While the tubes are moving upwards between the plates, let their velocity be $v$. One tube of force is supposed to proceed from each unit charge, so that the number of tubes per unit area of each plate is equal to $\sigma$, the surface density of electrification over it.

Now let us consider the magnetic force due to the two current-sheets which pass up one plate and down the other. By an investigation similar to that adopted in the case of the solenoid on page 103, it follows that the magnetic force is $4\pi c_1$, where $c_1$ is the current across unit length normal to the line of flow. The magnetic force, it will be seen, acts in a direction normal to lines of current-flow and normal to the electric tubes of force,—that is, in a direction perpendicular to the plane of the paper in the figure.

But the current across unit length is measured by the product of the charge per unit area into the velocity with which the charge moves. Thus $c_1$ may be replaced by $\sigma v$, and the magnetic force between the plates is $4\pi\sigma v$.

As shown on page 38,

$$4\pi\sigma = fk,$$

where $f$ is the electric intensity, and $k$ the dielectric constant. We thus have, for the magnetic force,

$$H = 4\pi\sigma v = fkv,$$

and, for the magnetic induction,

$$B = fvk\mu.$$

Now the essence of Maxwell's theory consists in the recognition of the magnetic effects of dielectric currents: that is, the recognition that a magnetic force is produced by a change in the dielectric polarization, a change which is equivalent to a motion of Faraday's tubes of electrostatic force. Let us, then, dismiss all ideas of conductors, and of ordinary electric currents in them, and fix our attention on a region of the dielectric medium through which are passing a succession of Faraday's tubes. According to the theory of Faraday and Maxwell, the existence of a magnetic force means the motion of electric tubes, and the magnetic force is $4\pi$ times the product of the number of tubes per unit cross area, a number which gives us $\sigma$ in the above equations, and of the velocity of the tubes at right angles to their length. The direction of the magnetic force is at right angles both to the length of the electric tubes, and to the direction of their motion.

Let us now return to the consideration of the propagation of a disturbance through the electromagnetic field, and approach the question from another side. Let us imagine that part of a circuit carrying a long straight current is moved in a magnetic field at right angles to its length and at right angles also to the direction of magnetic induction $B$. The rate of change of induction through the circuit is $Blv'$, where $l$ is the length of the moving wire, $v'$ its velocity, and $lv'$ the area swept out per second by the moving wire. Hence the electromotive force in the wire, or the difference of potential set up between its ends, is $Blv'$. Exactly the same effect is produced if we suppose the wire to be at rest, and the magnetic induction to be propagated through the field with the same velocity $v'$.

Once more let us suppose that we eliminate from our minds the idea of a conductor, and imagine the equivalent processes to occur in the field alone. The current is replaced by a change in the dielectric polarization, that is, by a movement of the electric tubes of force. Along the length $l$ of a line perpendicular to the direction of the motion there exists an electromotive force $E$, and thus at each point there is an electric intensity

$$f = \frac{E}{l} = Bv'.$$

But we have shown already that the motion of the electric tubes of force sets up a magnetic induction at right angles both to their length and to their direction of motion. Thus, in the direction of motion of the electric tubes, a wave of magnetic induction is propagated with the same velocity. This combined electric and magnetic disturbance represents an electromagnetic wave, and, putting $v$ equal to $v'$, we obtain a second relation between the quantities from the equation

$$f = Bv.$$

We now have two equations for $B$,

$$B = fvk\mu$$

and

$$B = \frac{f}{v}.$$

Therefore,

$$v^2 = \frac{1}{k\mu}, \text{ and } v = \frac{1}{\sqrt{k\mu}},$$

and this result, on Maxwell's theory, gives the velocity of an electro-magnetic disturbance propagated through an isotropic medium possessing a dielectric constant $k$ and a magnetic permeability $\mu$.

**68.** In the chapter on electrical units we saw that the dimensions of the electrostatic units, as ordinarily defined, were different from those of the usual electro-magnetic units for corresponding quantities. We also found that the ratio of any pair of units involved a velocity or some power of a velocity, and that the velocity in question was given by $1/\sqrt{k\mu}$. Thus it follows that, by comparing the numerical values of some one pair of units, we may determine the velocity with which an electromagnetic disturbance is propagated through the medium.

Much experimental skill has been devoted to the measurement of the number of electrostatic units in one electromagnetic unit of some electrical quantity. Weber and Kohlrausch found that one electromagnetic unit of quantity of electricity contained $3.1074 \times 10^{10}$ electrostatic units. Lord Kelvin compared the two units of potential, and found a value for $v$ of $2.93 \times 10^{10}$ centimetres per second. Clerk Maxwell balanced a force of electrostatic attraction against one of electromagnetic repulsion, and obtained $2.88 \times 10^{10}$.

On the whole, however, the most convenient pair of units to deal with are those of capacity; and experiments have been made by Ayrton and Perry, J. J. Thomson and Searle, and others, by this method. The electrostatic capacity of a condenser of regular form may, as we have seen, be calculated in simple cases from a knowledge of its form and dimensions. The same quantity, or the capacity of another condenser which may be compared with the first one, may be determined in electromagnetic measure in one of several ways. The most obvious method consists in charging the condenser to some known potential-difference by means of a standard voltaic cell, and then discharging it through a ballistic galvanometer, of which the constant is also determined. By this method Ayrton and Perry estimated $v$ to be $2.980 \times 10^{10}$ centi-

The ratio of the electro-static and the electro-magnetic units.

metres per second. Another arrangement, suggested by Maxwell and used by J. J. Thomson and Searle, consists in placing a condenser in one arm of a Wheatstone's bridge, and continually charging and discharging it by means of a revolving commutator (Fig. 92). A series of electric charges is thus conveyed through the arm of the bridge, and, if the period of the commutator is very short compared with the swing of the needle of the galvanometer, the discontinuous electric transfer

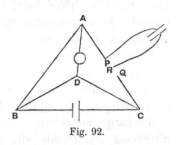

Fig. 92.

through the condenser-arm is equivalent to a continuous current. The charge conveyed by each vibration of the fork is $CV$, where $C$ is the capacity of the condenser, and $V$ the difference of potential between its plates. If $n$ be the number of complete vibrations of the fork per second, or the number of charges per second, the equivalent steady current is $nCV$. Now a steady current would be given by $V/R$, where $R$ is the resistance. Hence, the effective resistance of the arm of the bridge in which the condenser is inserted is $1/nC$. From such considerations it is possible to calculate the capacity of the condenser by applying Kirchhoff's laws to the bridge. Thomson and Searle used two coaxial cylinders, and calculated the capacity of the system in electrostatic units. They found that $v$ was $2\cdot9958 \times 10^{10}$ centimetres per second.

The mean of the determinations of the quantity $v$ is thus very near the value $3 \times 10^{10}$ centimetres per second. Now, within the narrow limits of experimental error, the observed velocity of light is also $3 \times 10^{10}$ centimetres per second. In Maxwell's words, "we have strong reasons for believing that light is an electromagnetic phenomenon."

**69.** Another consequence of the theory may be submitted to experimental examination. The velocity of an electromagnetic wave through air should bear to its velocity through some other transparent medium the ratio $1 : 1/\sqrt{k\mu}$ or $\sqrt{k\mu} : 1$. The magnetic permeability of ordinary transparent media is sensibly equal to that of air, and

*The relation between dielectric constants and optical refractive indices.*

this ratio becomes $\sqrt{k}:1$. But the ratio of the velocity of light in air to its velocity in another medium measures, on the undulatory theory and in practice, the refractive index $r$ of that medium. We should look, then, for a connexion between the optical index of refraction and the dielectric constant of different transparent substances; the theory indicates that the refractive index should be equal to the square root of the specific inductive capacity.

The specific inductive capacities tabulated on page 30 are obtained by measuring the relative capacities of condensers when their charges are retained for some considerable time—a time, at all events, comparable with a second. But the refractive index depends on the properties of the medium with respect to the exceedingly rapid vibrations of light, the periods of which are of the order $2 \times 10^{-15}$ seconds. The quantities are not strictly comparable. An attempt to secure comparable results has been made by determining the refractive indices for a series of vibration frequencies, and estimating their values for a frequency corresponding with an infinite wave length, by some such equation as that of Cauchy. But there is some doubt as to the accuracy of the values so obtained. Cauchy's formula does not contemplate any anomalous dispersion, and Maxwell's equations take no account of dispersion at all. Any comparison by such methods of the dielectric constants and the refractive indices still remains unsatisfactory.

| Substance | $\sqrt{k}$ | $r$ |
|---|---|---|
| Paraffin ... ... | 1·51 | 1·42 |
| Benzene ... ... | 1·54 | 1·50 |
| Carbon bisulphide ... | 1·63 | 1·64 |
| Flint glass ... ... | 3·18 | 1·71 |
| Quartz ... ... | 2·13 | 1·55 |
| Distilled water ... | 8·72 | 1·33 |

Thus, with certain substances, great discrepancies exist, though the considerations we have advanced above more than explain any such want of concordance. We are unable to produce electric alternations of periods comparable with those of light, but the electrical vibrations of systems of small capacity have frequencies rising to some hundreds of millions per second, and use has been made of these vibrations to determine some dielectric constants.

As we have seen, electromagnetic disturbances travel with a velocity which, in non-magnetic media, varies inversely as the square root of the dielectric constant. By observing the interference between two parts of an electromagnetic wave which have passed over paths of equal length in air and some other medium respectively, comparisons of velocity have been made. M. Blondlot and Sir J. J. Thomson independently found that, in rapidly changing fields, the specific inductive capacity of glass was less than in steady fields, and had a value of about 2·8. This gives $\sqrt{k}$ a value of 1·7, in agreement with the optical measurements of refractive index.

**70.** Clerk Maxwell published his electromagnetic theory of light in the year 1865. For twenty years the direct evidence in its favour was almost confined to the considerations we have given—the concordance of the value of the velocity involved in the ratio of the electrical units with the observed speed of light, and the agreement, more or less exact, between the indices of refraction of transparent media and the square roots of their dielectric constants. Maxwell's theory, although generally accepted in England, was but imperfectly known on the Continent, and its merits as a successful means of representing and co-ordinating electrical phenomena were hardly realized.

*The experiments of Hertz.*

In 1888, the physical world was startled by the announcement that electromagnetic waves had been produced, and their passage through space demonstrated, by Professor Heinrich Hertz, then of the Carlsruhe Polytechnic, who had found moreover that the waves moved with a speed which, at all events, did not differ much from the velocity demanded by Maxwell's theory—the velocity of light.

In the chapter on electromagnetic induction, it was shown that, on the analogy between self-induction and inertia, the discharge of an electrified system of capacity $C$ through a circuit of small resistance and of self-induction $L$, gave rise to electrical oscillations with a period of $2\pi\sqrt{LC}$.

One form of electric oscillator used by Hertz consisted of two metallic plates in the shape of squares with sides 40 centimetres in length. The plates were connected by rods about 30 centi-
metres long with two small highly
polished gilt balls placed 2 or
3 centimetres apart. The plates
were connected with the opposite

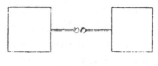

Fig. 93.

terminals of the secondary circuit of an induction coil. Each time the plates became charged to a certain difference of potential a spark was produced, and electric oscillations were set up. With the apparatus used by Hertz, the period of alternation was about $1{\cdot}85 \times 10^{-8}$ seconds. The resistance was high and the self-induction low; consequently very few oscillations occurred at each discharge, the energy being partly radiated as waves, and partly dissipated as heat in the spark.

To detect the presence of the electromagnetic waves in the space surrounding the oscillator, Hertz made use of the principle of resonance. When any physical system, possessing a natural period of vibration of its own, is subjected to periodic impulses which coincide with the natural period of the system, violent oscillations are set up. This principle is illustrated by the timing of the impulses given to a child's swing, and by the sounding of a piano wire of one particular frequency of vibration when a note in unison with its own is sung near it.

Hertz's resonator was formed of a circle of wire about 35 centimetres in radius, separated at one part of its
circumference by a spark-gap made by two small
knobs. By means of a screw, working in an
ebonite frame, the knobs could be adjusted at
any small distance from each other. We may
regard an uncharged conductor as possessing

Fig. 94.

equal charges of the opposite kinds of electricity;
and, when properly timed impulses fall on it, electric surgings of these charges will be set up. As the positive charge swings one way and the negative the other, the knobs become oppositely electrified, and their potential-difference may become so great, owing to the kinetic energy of the oscillating charges, that a spark passes.

A circle of this size possesses a natural period of electric oscillation about equal to that of the oscillator used by Hertz, and thus, if it be moved about in front of the spark-gap of the oscillator, resonance effects may be observed by the presence of minute sparks, the length of which depends on the position of the plane of the resonator and on the position of the air-gap therein.

The electric waves given by Hertz's apparatus may be reflected by metal surfaces, focussed by parabolic mirrors of sheet zinc, and refracted by large prisms of pitch. A frame of parallel wires is transparent to the waves when the wires are perpendicular to the line of the spark, and opaque when the wires are parallel to the spark. In all these respects, the electromagnetic waves show the properties of plane polarized light, in which, according to the undulatory theory, the vibrations occur in straight lines in a plane at right angles to the direction of propagation.

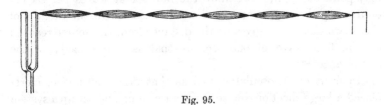

Fig. 95.

When the waves of the sea impinge directly on a straight wall, they are reflected. Interference results between the direct and reflected waves, and a system of what are called stationary undulations is set up. The motion of the water in front of the wall is such that no waves appear to be moving forward; any point on the water simply rises and falls periodically. At certain intervals this motion ceases, and we get points called nodes, where the water is stationary. Another illustration of the same principle is given by the transverse vibrations of a stretched string or wire. If both ends be fixed, the fundamental mode of vibration gives a node at each end, and an internode or loop at the middle. This arrangement may be considered to be due to the successive reflection of the waves at the ends of the string. If the string be held lightly at the centre, it may be made to vibrate in halves, and, if it is emitting a sound, the note will be the octave of the first.

Other modes of vibration, giving higher overtones, may also be produced, the string being more subdivided. The analogy between the vibrations of the string and the electric waves we have now to consider is best seen if we imagine the vibrations of the string to be set up by connecting one end to a tuning fork, the vibrations of which are maintained electrically, and keeping the other end of the string fixed. Waves pass out from the fork, and, if the length of the string be an exact multiple of the half wave-length of the undulations travelling along it, stationary waves are seen, owing to the persistence of visual impressions, the string being visibly divided into a series of nodes and loops. This system again may be regarded as due to the interference of the direct and reflected trains of waves, and from the figure it will be evident that the distance between two nodes is equal to half the complete wave-length of the undulations travelling along the string. We may point out that, if we know the number of waves per second emitted by the source, and the length of each wave, the product of these two quantities gives us the distance from the source reached by the first wave in one second—that is, the velocity of the disturbance.

In front of an oscillator such as that described above, Hertz placed a large sheet of zinc to act as a reflector, and set up a system of stationary electric undulations. When he explored the space between the oscillator and the reflector, he found that, at certain places, the length of the sparks obtained in his resonator was a maximum, while, at other intermediate places, the spark-length was a minimum. Thus we obtain by direct experiment the wave-length of electric oscillations of known period, and can deduce an approximate value for the velocity. Within the limits of experimental error, it proves to be the same as that of light. Maxwell's theory receives a direct confirmation, of the most striking kind.

Hertz used a resonator constructed to have as nearly as possible the same period of vibration as the oscillator, and the wave-lengths, as measured by him, are those of the undulations which synchronize both with the vibrations of the oscillator and of the resonator. In repeating his experiments with a series of resonators, Sarasin and de la Rive found that the wave-length indicated by the distance between the successive nodes was the wave-length of the vibrations of the resonator, not of the oscillator. The explanation of this result is found in the fact that a system like the oscillator is a very good radiator of electromagnetic energy, and soon loses its amplitude of oscillation. The amplitude of the tenth swing has been shown to be only about the 1/14 of the first. On the other hand, the resonator has been shown to be a very slow radiator; oscillations once set up in it maintain their energy for a considerable time, and more than a thousand swings are needed to reduce the amplitude to 1/10 of its original value. When the disturbance from the oscillator falls on the resonator, electric surgings are set up, even though the tuning be incomplete; the first few impulses being almost the only effective ones. The resonator goes on vibrating long after the train of waves from the oscillator has ceased; and, although its energy of radiation is much less, it is the direct and reflected waves from the resonator which form the persistent stationary wave-system investigated in the region in front of the reflector in the experiments of Hertz and in those of Sarasin and de la Rive.

We have seen that the energy of an electric current may be regarded as passing through the dielectric medium, the circuit in a wire being merely the line of dissipation of the energy into heat. But, just as a wire serves to direct the flow of energy in a steady current, so the energy of electromagnetic waves may be guided along wires. Here again, the waves pass through the surrounding medium, and must travel with the velocity of light in air, and with the velocity $1/\sqrt{\mu k}$ through other substances. While in the field surrounding a long straight steady current, Faraday's tubes will be parallel to the current, and will be moving into the wire at right angles to their length, with rapidly alternating disturbances, the electric tubes will be radial to the wire and will move backwards and forwards along the wire with one end slipping along in its substance.

**71.** The theory of Clerk Maxwell and the experiments of Hertz have borne practical fruit in their application
Wireless
telegraphy.    to the problems of telegraphy. The resonator used by Hertz is not a convenient means of detecting the incidence of electromagnetic waves passing through free space, and better

arrangements were necessary before technical use could be made of the new discoveries.

Sir Oliver Lodge, Branly and Marconi introduced a form of detector known as the coherer. If metallic filings (*e.g.* a mixture of nickel and silver) be placed in a small gap between two silver plugs in an exhausted glass tube, the electrical resistance between the plugs is, in the normal state of the tube, very great. If, however, electromagnetic radiation is falling on the tube, the resistance is much diminished, and a current will pass through a relay circuit including a battery, a galvanometer, and the coherer. If the coherer be kept constantly tapped, so as to shake the filings, it returns to its initial state of high resistance when the waves cease, and the current through it is stopped.

Receivers of more recent date are founded on the effect of electromagnetic waves in demagnetizing iron. If a piece of iron wire be drawn through a magnetic field, its magnetization persists owing to hysteresis. When electromagnetic waves are passed round a solenoid encircling the wire, the iron follows the magnetizing force more closely, the hysteresis effect is diminished, and the induction in a second coil changed. This is detected by a telephone. Instruments of this type have been used by Marconi.

Other receivers not only detect oscillations but also rectify them—that is, allow to pass currents in one direction only. The simplest example is the crystal detector. Certain crystals such as zincite, galina or carborundum, when pressed against a metal edge or point will act in this way, especially if a difference of potential of the order of a volt be kept up between the metal and the crystal.

Another rectifying detector is the therm-ionic valve. If a metal plate be fixed inside an electric lamp, a negative current carried by ions (see Chapter XI) will pass from the hot filament to the plate but not in the reverse direction. The waves may be detected by a galvanometer or by a telephone. In more recent forms of valve a cylindrical gauze surrounds the tungsten filament, and a metal cylinder surrounds the gauze. The resistance between cylinder and gauze depends on the direction of the electric field, for when the gauze is negative with respect to the filament, no electrons leave the filament. But if the filament be negative, electrons are emitted, pass through the gauze and can carry a current between it and the outer cylinder.

For long-distance signalling, large quantities of energy are necessary, and as much of that energy as possible must be collected at the receiving station. Hence it was only when Marconi introduced the use of an insulated vertical wire 100 or 150 feet in height to collect the radiation, that it became possible to telegraph over great distances. The arrangement of a simple form of apparatus for wireless telegraphy is shown in Figure 96. The induction coil

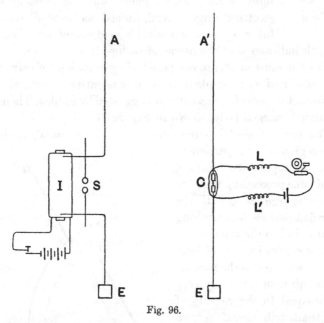

Fig. 96.

$I$ is connected with a spark-gap $S$, the opposite sides of which are joined, one to the earth, and the other to the long aerial wire $A$. At the receiving station, another aerial wire is connected with one terminal of the coherer, the other being put to earth. The coherer is also in a relay circuit, consisting of a battery, a bell, and two coils $L$, $L'$ possessing self-induction.

In more recent forms of apparatus the oscillations are induced in the aerial wire of the transmitting station by other oscillations in a primary circuit of larger capacity and energy-storing power. This plan increases greatly the effective range. The receiving station is also furnished with two circuits, and oscillations are induced in the receiver circuit by those set up in the sky-wire. By adjusting the periods of vibration of the four circuits, it has been found possible to prevent the responding of the receiver to other oscillations than those to which it is "in tune."

**72.** It is interesting to consider how a wave of light must
start and be propagated through the aether. If light be an electromagnetic wave, it must have an electrical origin. And, indeed, as we shall see in a future chapter, cumulative evidence of overwhelming

*The origin and mode of propagation of waves of light.*

strength indicates that the atoms of bodies, from the vibrations of which light must arise, are composed of aggregations of corpuscles each associated with, or identical with, a negative electric charge.

Now let us consider, in a manner suggested by Sir J. J. Thomson, an isolated charged body, shown in Figure 97.

The lines of electric force must evidently be radial, as shown in the region near the particle $O$, where $Op$ represents one such line of force proceeding from the electric charge at $O$. If the electrified particle be travelling forwards, in the direction of the arrow, it carries its lines of force with it; and, unless the particle be moving with a velocity very nearly equal to the velocity of light, the distribution of the lines is unaltered; they still are uniformly spaced radii, proceeding

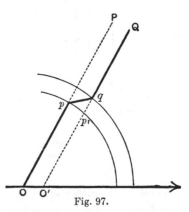

Fig. 97.

from the particle as centre. Whenever electric tubes of force are moving, there exists a magnetic force at right angles, both to their length and to their direction of motion, and therefore a magnetic field must be produced by the moving particle $O$.

Now let us imagine the moving particle to be stopped suddenly. If a change could be propagated instantaneously throughout all space, the lines of force would stop at once also. But a change in

electromagnetic properties can be propagated only with the speed
of an electromagnetic wave, that is, with the velocity of light.
Thus, when a moving electrified particle is arrested, a pulse of
electromagnetic force starts from the particle as its centre, and
spreads out in spheres, rectifying the distribution of the lines of
force as it goes. The effect is shown in Fig. 97. If the particle
had not been stopped, at the end of an interval of time, $t$, it would
have reached some new position $O'$, and the lines of force would
be radii from this point as centre. Beyond the sphere reached by
the rectifying pulse, the lines of force will still be moving parallel
to the direction of motion of $O$, and, at the instant considered, will
be radii of the point $O'$, while, behind the spherical pulse, the lines
will be at rest, and will be radii of the point at which the particle
is stopped. The lines of force must be continuous; and therefore,
in the pulse itself, the lines must run in some direction such as $pq$
in the figure. The electric force near $pq$ has then a component at
right angles to the direction of propagation of the disturbance, that
is, at right angles to the radial lines. Whenever a Faraday tube
of electric force moves, it produces a magnetic force at right angles
both to its length and to its direction of motion, and thus the line
of force $pq$ within the pulse produces a magnetic force at right
angles to the plane of the paper. Now the waves of light, and, to
pass to much greater wave-lengths, the waves used in wireless
telegraphy, are aethereal waves of electromagnetic force so
arranged that the electric and magnetic forces are at right angles
to each other, and both at right angles to the direction of propagation of the waves. It follows that the pulse, indicated by our figure
as spreading out, consequent on its arrest, from a moving electrified
particle, is a pulse of the same nature as the waves of light, with
this exception, that, instead of a series of regular periodic waves,
it consists of a single expanding shell of electromagnetic force.

But, if, instead of imagining the moving corpuscle suddenly
brought to rest, we suppose that it is reversed in its path, and that
this reversal occurs periodically, so that the corpuscle performs
simple harmonic vibrations, we get, instead of a single thin pulse, a
series of less abrupt but regularly recurring alternations propagated
out from the corpuscle as centre. Each Faraday's tube is set into
oscillation at its inner end, and transverse waves travel outwards

along it, just as waves travel along a stretched cord, when one end is oscillated periodically by the hand. The distribution of electric and magnetic forces in the advancing wave-front is essentially the same as in the case of the sudden pulse already studied: we get, in fact, a series of regular aethereal waves, in which there are electric and magnetic forces, both in the plane of the wave-front and at right angles to each other in that plane. But such an arrangement is precisely that required to explain the phenomena of light.

In the simple case we have taken, the corpuscle oscillates backwards and forwards in a straight path: the vibrations travel as tremors along the tubes of force in one plane only; the resultant light is plane polarized. In the more general case, we must suppose that the corpuscle oscillates in a circular, or elliptical orbit, and the tubes of force will be displaced in corresponding motions; the tremors running along them will no longer be simple to and fro movements; points on the tubes will describe curved paths. These paths change continually as the orbit of the corpuscle changes, and we get a complete model of the propagation of common, non-polarized light.

Faraday's tubes, it is clear, give a very powerful and convenient method of studying the phenomena of the electromagnetic field, and indications are not wanting that they represent something more than a useful mathematical fiction. If the structure of the electric field be discontinuous in reality, as our tube-picture of it indicates; if the electric and magnetic effects of a charge of electricity are in reality exerted throughout the surrounding space by means of discrete tubes of force—vortex filaments in the aether, or whatever they may actually be—an advancing wave of light must be discontinuous also. Could we look at such a wave from the front, and magnify it millions of millions of times, we should see, not a uniform field of illumination, but a number of bright specks scattered over a dark ground. Each tube of force would convey its own tremors, and these would constitute light, but between them would lie undisturbed seas of aether.

Such an idea about the nature of a wave-front of light is very unexpected and surprising. We are inclined at once to relegate our tubes of force to a museum of conceptual curiosities. But it is a remarkable thing that certain evidence in favour of the discontinuous nature of a wave-front of light really does exist. This evidence depends on electric conduction through gases, the phenomena of which will be described in a future chapter.

**73.** The laws of force are, as we have seen, of the same form for magnetism as for electrostatics, when pole-strength replaces charge, and magnetic permeability is put instead of specific inductive capacity. The energy of the electrostatic field (§ 23) is $f^2k/8\pi$ per unit volume, where $f$ denotes the electric intensity; and it must follow, by a similar method of proof, that the energy per unit volume of a magnetic field is $H^2\mu/8\pi$, where $H$ is the magnetic force.

*The energy of the electro-magnetic field.*

Now, in terms of Faraday's tubes of electric force, we explain a magnetic force as due to the motion of the electric tubes at right angles to their length, and thus, if the tubes move with a velocity $v$, we have, as in § 67,

$$H = 4\pi Nv,$$

when the number of tubes per unit cross area is $N$, and the tubes move at right angles to their length. If the direction of their motion makes an angle $\theta$ with their length,

$$H = 4\pi Nv \sin \theta.$$

We may now write the magnetic energy per unit volume of the field in terms of the number and velocity of Faraday's tubes of electrostatic induction. The magnetic energy has the value

$$\frac{H^2\mu}{8\pi} = \frac{\mu}{8\pi}(4\pi Nv \sin \theta)^2$$

$$= \tfrac{1}{2} 4\pi\mu N^2 v^2 \sin^2 \theta.$$

If we regard this energy as the kinetic energy of the tubes, and the expression for the energy as analogous to the familiar formula $\tfrac{1}{2}mv^2$, we see that the energy is that of a mass $4\pi\mu N^2$ moving with the velocity of the tubes in a direction at right angles to their length.

The kinetic energy of a freely moving body does not depend on the direction of motion, but a dynamical analogy with the magnetic energy of the tubes is found in the kinetic energy of a cylinder moving through a liquid.

A body travelling through a liquid disturbs some of the liquid through which it passes, and thus the effective mass of the body is increased. If the body be a sphere, the added mass is evidently independent of the direction of motion; but, if a cylinder be made to travel, it will disturb more liquid when going sideways than when proceeding in the direction of its length. If its length be very great compared with its breadth, it will, when moving in the direction of its length, simply glide through the liquid, and if that liquid possesses no viscosity, none of it will be set in motion by the cylinder. Moving sideways the cylinder will disturb a certain mass $M$ of the liquid, and moving at an angle $\theta$ with its length, it will produce momentum $Mv \sin \theta$ at right angles to its length, the energy due to the motion being $\frac{1}{2} Mv^2 \sin^2 \theta$.

Our analogy is now clear. The magnetic energy of the field is $\frac{1}{2} 4\pi\mu N^2 v^2 \sin^2 \theta$; and thus we may say that the energy is the same as the kinetic energy which would exist if real tubes moved through a material frictionless liquid, and moved a mass of that liquid, the total mass moved being equal to $4\pi\mu N^2$ when the tubes travel at right angles to their length.

The mass of aether bound to each tube is $4\pi\mu N$, and therefore depends on the number of tubes per unit cross area. This result increases the completeness of the analogy with a moving cylinder, for a system of many cylinders fixed together would set in motion more liquid than would the same number of cylinders if moving independently. Nevertheless, the analogy is only traced as giving a rough idea how the electromagnetic equations may represent a dynamical possibility. It is not essential to the course of the investigation now to be resumed.

The electrostatic energy per unit volume of the field is $f^2 k/8\pi$, and, as we saw in § 22, the number $N$ of Faraday's tubes per unit area, which measures the electrostatic polarization, is $fk/4\pi$. Hence the potential energy per unit volume is $fN/2$, or, since $f$ is $4\pi N/k$, the energy is $2\pi N^2/k$.

The mass $M$ which we suppose moving is $4\pi\mu N^2$, and therefore the potential energy per unit volume is

$$\tfrac{1}{2} M \frac{1}{\mu k} = \tfrac{1}{2} M V^2,$$

where $V$ is the velocity of light in the medium. Hence the electrostatic potential energy of the field is equal to the kinetic energy which the moving mass would possess if travelling with the velocity of light.

The energy of a charged condenser is the potential energy of electrostatics. When the condenser is discharged by means of a wire, the statical energy changes suddenly into the kinetic energy of moving Faraday's tubes, and, in the end, some of this energy is converted into heat in the conducting wire. Meanwhile, oscillations may be set up, alternations of potential and kinetic energy occur, and electromagnetic waves be emitted.

In wave motion, we are accustomed to look for changes in energy from the kinetic to the potential form; hence, on our present analogy, we should regard the energy per unit volume at a fixed point when a train of electromagnetic waves passes over it as alternating from the magnetic to the electric form.' When all the energy is kinetic, its value per unit volume is $H^2\mu/8\pi$, and when potential, $f^2k/8\pi$.

**74.** If we refer the energy of the magnetic field to the kinetic energy of moving tubes of electric force, it follows that the field must contain momentum—the momentum, in fact, of the tubes and the aether which we may imagine to be disturbed when they move.

The energy of the magnetic field is $\frac{1}{2}4\pi\mu N^2v^2\sin^2\theta$, with the same notation as above, and the effective mass $M$ of the tube and its attendant aether is $4\pi\mu N^2$. The momentum $Mv$ is thus $4\pi\mu N^2v\sin\theta$, or, if we take the simplest case, when the tubes move at right angles to their length, $4\pi\mu N^2v$. Since the magnetic force $H$ is $4\pi Nv$, we may write this momentum as $\mu HN$ or $BP$, where $B$ is the magnetic induction and $P$ the electrostatic polarization of the medium.

Now it is important to understand that this aethereal momentum has an existence quite as real as the momentum of an ordinary moving body. Momentum can pass freely from aether to matter, and from matter back again to aether. If this were not so, we could not extend Newton's laws of motion to a system consisting of a charged body and a medium through which passes an electric pulse. The body would be set in motion, and, unless we take into account the momentum in the medium, action and reaction would not be equal and opposite.

Another important case is that of an electromagnetic wave falling normally on an absorbing surface. Here we have energy alternating between the kinetic and potential forms, and, in any given volume of the medium, we may take half the energy at any instant to be the kinetic energy of moving tubes. The energy per unit volume of the medium is $\frac{1}{2}4\pi\mu N^2v^2$, and, on the average, half of this is kinetic. The average momentum, then, is half that corresponding with steady motion, or $\frac{1}{2}4\pi\mu N^2v$ per unit volume.

The radiation in $v$ units of volume of the medium reaches unit area of the receiving surface in one second. Hence, the momentum communicated to unit area of the surface per second, that is, the pressure on the surface, is $\frac{1}{2}4\pi\mu N^2v^2$ which, it will be noted, is equal to the total energy, kinetic and potential, contained in unit volume of the medium.

We have supposed the surface to absorb the radiation. If we use a reflector, the momentum is returned to the medium in the reverse direction, so that the change of momentum is doubled, and the pressure of radiation is doubled with it.

A body exposed to electromagnetic radiation of any kind, including light, will be pushed in the direction of the incident beam by a pressure which is calculable if we know the amount of energy received per second and the velocity of light. For instance, the heat received per square centimetre by surfaces exposed normally to bright sunlight on the earth is about 2 thermal (gramme-degree-centigrade) units per minute, or 1/30 heat unit per second. In dynamical units of energy this is equal to $4 \cdot 2 \times 10^7/30$, or $1 \cdot 4 \times 10^6$ ergs per second. This quantity of energy, received on one square centimetre in one second, is, at the beginning of the second, spread out through space, and occupies a column one square centimetre in cross section, and 186,000 miles, or $3 \times 10^{10}$ centimetres, in length. The energy per unit volume, then, is, at any instant,

$$\frac{1 \cdot 4 \times 10^6}{3 \times 10^{10}} = 4 \cdot 7 \times 10^{-5} \text{ ergs per cubic cm.}$$

This number must also represent the force per unit area, in dynes per square centimetre.  Hence, on one square metre of an absorbing surface, exposed to bright sunlight, the force is

$$0{\cdot}47 \text{ dyne} = 0{\cdot}46 \text{ milligram-weight.}$$

This pressure, which we have calculated from the conception of the momentum of moving tubes of force, was originally deduced by Maxwell as a consequence of the electromagnetic theory.  Larmor has shown that it is necessary on any theory of undulations, and Bartoli has deduced it on thermodynamical principles by an application of Carnot's idea of the reversible cycles of a heat engine.

As we have seen, on a reflecting surface the pressure is twice as great as on an absorbing surface, and, by seeking for the difference between the forces on the absorbing and reflecting surfaces of suspended vanes, Professor Lebedef of Moscow has demonstrated experimentally the existence of this minute radiation-pressure, and his results have been confirmed and extended by Professors Nichols and Hull in America, while Professor Poynting has detected the tangential effect of light incident obliquely.

**75.**  An electrostatic charge is the origin of Faraday's tubes
Magnetic forces as due to the motion of electrostatic tubes of force.
of electric force, and thus, on Maxwell's theory, the motion of a charged conductor should produce a magnetic field like that due to a current.
This conclusion has been confirmed experimentally by Rowland, Pender, and others.  Two circular plates, divided into sectors by insulating divisions, are fixed parallel to each other, on one axis, which passes through their centres and is normal to their planes.  When the sectors are similarly electrified and set in rotation, a magnetic force is produced, equivalent to that which would be caused by circular currents flowing in the direction of motion of the electric charges.  The magnetic field can be detected by the deflection of a needle suspended between the plates.

R. W. Wood has modified this investigation by using a flight of electrified particles of solid carbon dioxide.  An iron cylinder containing the liquefied gas becomes electrified negatively when the gas is allowed to escape through an orifice.  Some of the gas solidifies owing to the cold of expansion, and the positively electrified particles are allowed to fly along a glass tube placed under a magnetic needle. A deflection of the needle results which may be reversed by reversing the direction of the jet.

A permanent steel magnet is accompanied by a steady magnetic field in its neighbourhood. On our present theory, that field must be due to the motion of Faraday's tubes of electrostatic force, and hence the magnet must be the centre of a system of moving electric tubes. The work done in magnetizing a steel bar must include the energy needed to set the tubes in motion, and, once established, the motion must be regarded as inseparable from the magnet itself. If we move the magnet, we move with it its attendant system of moving electric tubes.

In a steady magnetic field there are no electromotive forces—no induced currents are found unless the magnetic field is changing. But a single tube of electrostatic force implies an electric intensity, which tends to move electricity, and thus to set up a current. We are driven to conclude, therefore, that, in a steady magnetic field, negative as well as positive tubes of electric force must exist in motion, and that as many negative as positive tubes enter any element of volume per second.

Another aspect of this problem is seen if we remember that positive electric charge induces negative charge on a metal plate in its neighbourhood. If the positive charge be set in motion, it produces a magnetic force, which acts through the metal plate unless that plate be made of thick iron. On the usual view of electrostatic tubes of force, we should imagine them as springing from the positive charge, and ending on the induced negative charge, and it is difficult to see how their motion could then cause a magnetic field beyond the metal screen. For reasons such as these, Sir J. J. Thomson has imagined both positive and negative electrostatic tubes of force, starting respectively from positive and negative units of charge, and running from these charges out into space. If we suppose that the nature of the electric tubes may be represented mentally by some such idea as that of vortex filaments in the aether, it is easy to picture positive and negative tubes as oppositely circulating vortices.

These tubes will attract each other if of opposite kind, and will repel each other, as Faraday supposed, if of the same kind. As far as electrostatic action is concerned, Thomson's conception is equivalent to the older ideas, but it has the advantage of being applicable also to electrodynamic phenomena. The exact meaning of positive and negative electricity remains, however, uncertain. In a future chapter we shall return to the question of the relation between them.

## REFERENCES.

*Elements of the Mathematical Theory of Electricity and Magnetism*; by J. J. Thomson; Chapter XIII.

*Electricity and Magnetism*; by J. Clerk Maxwell; Chapters XIX, XX, XXI.

*Electricity and Matter*; by J. J. Thomson.

*Encyclopædia Britannica*, Supplement; Arts. "Electric Waves," "Telegraphy."

# CHAPTER X

## ELECTROLYSIS

**76.** In Chapter V on the Electric Current we described the origin of the voltaic cell and the change it effected in the direction of the main stream of electrical research. The attention of physicists, who had been occupied with the phenomena of electrostatics and of the disruptive discharge of electric machines and Leyden jars, was turned to the new field of inquiry, and the opening years of the nineteenth century witnessed a rapid development of knowledge, particularly in the chemical effects of the so-called galvanic or voltaic currents.

The fundamental observation, from which arose the science of electro-chemistry, was made in the year 1800, immediately on the news of Volta's discovery reaching England.

Using a copy of Volta's original pile, Nicholson and Carlisle found that when two brass wires leading from its terminals were immersed near each other in water, there was an evolution of hydrogen gas from one, while the other became oxidised. If platinum or gold wires were used, no oxidation occurred, but oxygen was evolved as gas. They noticed that the volume of hydrogen was about double that of oxygen, and, since this is the proportion in which these elements are contained in water, they explained the phenomenon as a decomposition of water. They also noticed that a similar kind of chemical action went on in the pile itself, or in the cups when that arrangement was used.

Soon afterwards, Cruickshank decomposed the chlorides of magnesia, soda and ammonia, and precipitated silver and copper from their solutions—an observation which afterwards led to the process of electroplating. He also found that the liquid round the pole connected with the positive terminal of the pile became acid and the liquid round the other pole alkaline.

In 1806 Sir Humphry Davy proved that the formation of the acid and alkali was due to impurities in the water. He had previously shown that decomposition of water could be effected although the two poles were placed in separate vessels connected together by vegetable or animal substances, and established an intimate connexion between the galvanic effects and the chemical changes going on in the pile.

The identity of " galvanism " and electricity, which had been maintained by Volta, and had formed the subject of many investigations, was established in 1801 by Wollaston, who showed that the same effects were produced by both, while in 1802 Erman measured with an electroscope the potential differences furnished by a voltaic pile.

In 1804 Hisinger and Berzelius stated that neutral salt solutions could be decomposed by electricity, the acid appearing at one pole and the metal at the other, and drew the conclusion that nascent hydrogen was not, as had been supposed, the cause of the separation of metals from their solutions. Many of the metals then known were thus prepared, and in 1807 Davy decomposed potash and soda, which had been considered to be elements, by passing the current from a powerful battery through them when in a moistened condition, and so isolated the metals potassium and sodium.

The decomposition of chemical compounds by electrical means indicated a connexion between chemical and electrical forces. Davy " advanced the hypothesis that chemical and electrical attractions were produced by the same cause, acting in one case on particles, in the other on masses." This idea was developed by

Berzelius, who regarded every compound as formed by the union of two oppositely electrified parts—atoms or groups of atoms. The exact dualistic formulation of his theory given by Berzelius was afterwards abandoned, but the essence of the idea—the explanation of chemical forces in terms of electrical conceptions—remains to this present.

The remarkable fact that the products of decomposition appear only at the poles was perceived by the early experimenters on the subject, who suggested various explanations. Grotthus in 1806 supposed that it was due to successive decompositions and recombinations in the substance of the liquid. Thus, if we have a compound $AB$ in solution, the molecule next the positive pole is decomposed, the $B$ atom being set free. The $A$ atom attacks the next molecule, seizing the $B$ atom and separating it from its partner, which attacks the next molecule and so on. The last molecule in the chain gives up its $B$ atom to the $A$ atom separated from the last molecule but one, and liberates its $A$ atom at the negative pole.

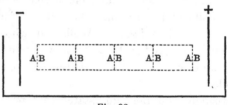

Fig. 98.

A new terminology, which is still used, was introduced by Faraday in 1833. Instead of the word *pole*, which implied the old idea of attraction and repulsion, he used the word *electrode*, and called the plate of higher electric potential, by which the current is usually said to enter the liquid, the *anode*, and that by which it leaves the liquid, the *cathode*. The parts of the compound which travel in opposite directions through the solution he called *ions*—*cations* if they went towards the cathode, and *anions* if they went towards the anode. He also introduced the words electrolyte, electrolyse, etc., which we have already used.

Faraday pointed out that the difference between the effects of a frictional electric machine and of a voltaic battery lay in the fact that the machine produced a very great difference of potential, but could only supply a small quantity of electricity, while the battery gave a constant supply, much larger in quantity, but only produced a very small difference of potential.

**77.** If we connect together in series a single Daniell's cell, a galvanometer, and two platinum electrodes dipping into acidulated water, no visible chemical decomposition ensues. At first a considerable current will be indicated by the galvanometer; the deflection soon diminishes, however, and finally becomes very small, though this small current seems to leak through permanently.

Polarization.

If, instead of using a single Daniell's cell, we employ some source of electromotive force which can be varied as we please, and gradually raise its intensity, we shall find, when it exceeds a certain value, about 1·7 volts, that a permanent current of considerable strength flows through the solution, and, after the initial period, shows no signs of decrease. This current is accompanied by chemical decomposition.

Now let us disconnect the platinum plates from the battery, and join them directly with the galvanometer. A current will flow for a while in the reverse direction ; the system of plates and acidulated water, through which a current has been passed, acts as an accumulator, and will itself yield a current in return.

These phenomena are explained by the existence of a reverse electromotive force at the surfaces of the platinum plates. Only when the applied electromotive force exceeds this reverse force of polarization, will a permanent steady current pass through the liquid, and visible chemical decomposition proceed.

Recent experiments, by Le Blanc, Oberbeck, and others, have shown that the reverse electromotive force of polarization is due to the deposit on the electrodes of minute quantities of the products of chemical decomposition. Differences between the two electrodes are thus set up, and, if disconnected from the external electromotive force, the arrangement would act as a source of current, just as does a primary cell. When acted on by the external electromotive force, the effect of the deposits is to oppose a reverse electromotive force to that in the external circuit. As the primary current continues to pass, the densities of the deposits increase, and with them grows the reverse electromotive force, till a continuous film of deposit is formed. This film has, by calculation from the total electric transfer needed to produce it, a thickness of the order of $10^{-8}$ centimetre—a new measurement of the thickness of a layer of molecular dimensions. The reverse electromotive force of polarization then reaches its maximum value. If the primary external electromotive force be not as great as this maximum reverse force, the maximum value is, of course, not reached; the polarization grows till the reverse electromotive force is sensibly equal to that applied, and the current nearly stops. The slight leakage current which remains is probably due to the gradual diffusion away from the electrodes of the products of the decomposition. If the applied electromotive force be greater than the maximum force of polarization, a permanent current flows, but the effective electromotive force of the circuit is only the excess of the applied force $E$ over the reverse force $E'$, the current being $(E - E')/R$.

In the case we have chosen, hydrogen and oxygen are evolved from acidulated water at the surfaces of bright platinum electrodes. In contact with these two gases respectively, the system will act as an accumulator, and continue to give a reverse current till the gases are exhausted. Now the maximum reverse force of polarization, about 1·7 volts, is greater than the electromotive force which the polarized plates themselves exhibit when used as a source of current. It seems that the process is not reversible in the thermodynamic sense of the word. If, however, the electrodes be covered by a deposit of platinum black, by previously passing a current backwards and forwards between them through a solution of platinum chloride, Le Blanc has shown that the minimum decomposition point is 1·07 volts, which is equal to the electromotive force of the oxyhydrogen gas battery. In this case, then, the process is reversible. If an external electromotive force of 1·07 volts be applied, the system is in equilibrium; while, if the applied force exceeds or falls short of that value by an infinitesimal amount, an indefinitely small current will pass one way or the other, and the gases are slowly evolved or absorbed. The platinum black has a very large surface, and, owing to the well-known occluding power of platinum, the gases are probably absorbed in the electrodes as fast as they are produced. They diffuse through the substance of the electrodes, and dissolve in the liquid or escape into the atmosphere. On the other hand, when bright plates are used, the exposed surface is too small to absorb the gases, which must therefore be evolved directly as bubbles at the plates. In this process a certain amount of irreversible work is done, and the applied electromotive force rises to 1·7 volts before it can overcome the opposing force.

By methods we shall descril э later, it is possible to separate the potential-differences at the aɪ ɪde and the cathode, and, although some doubt remains as to the trustworthiness of the absolute values, the results may fairly ѣ ɪ used when absolute values are not necessary. It has been concluded from several series of experiments that, in the case of all substances examined, the deposition and solution of metals in contact with solutions of their salts are reversible processes—the decomposition voltage is equal to the reverse electromotive force which the metal itself gives when going into solution.

**78.** During the early investigation of the subject, it was
The nature of the ions. thought that, since hydrogen and oxygen were usually evolved, the electrolysis of solutions of acids and alkalies was to be explained as a direct decomposition of water, the function of the acid or alkali being imagined simply to be to give conductivity to the otherwise non-conducting solvent. When salt solutions are examined, other substances, such as metals, are deposited at the electrodes, and it is necessary to suppose that the solute itself takes part in the process of conduction. During the electrolysis of a solution of copper sulphate, copper is deposited at the cathode, and copper is dissolved from a copper anode, or oxygen and sulphuric acid liberated if the anode be of platinum. These phenomena are explained readily by the hypothesis that the ions are a positively electrified copper atom, which may be written as $Cu+$, and the negatively electrified acid group $SO_4-$, which combines with copper, or, if no dissolvable metal be present, attacks water to form $H_2SO_4$ and oxygen. No facts are known in this case inconsistent with such a view of the process; but, in solutions in some other solvents, a similar supposition seems insufficient. It is well to remember that the simple case of copper

sulphate is also explicable on the theory that the ions result from
the dissociation of a complex molecule formed by the combination
of the salt and the water. The ions $\overset{+}{Cu}(H_2O)$ and $\overset{-}{SO_4}(H_2O)$ would
produce exactly the same phenomena as those to be expected from
the action of the simpler structures, and the same effects would
result if the charged particles were associated with a number of
water molecules.

A study of the products of decomposition alone does not lead
necessarily to a knowledge of the ions involved in the passage of
the current through the electrolyte. The electric force is active
throughout the whole solution; all the ions must come under its
influence and therefore move, but some may need a smaller electromotive force than others for their evolution at the electrode, and
consequently, as long as any quantity of all the ions of the solution
remains in the layer of liquid next the electrode, only these ions
will be evolved.

The issue is obscured in another way also. When the ions are
set free at the electrodes, they may unite with the substance of the
electrode or some constituent of the solution and form secondary
products. For instance, there is reason to suppose that, in a dilute
solution of sulphuric acid, the ions either are or contain hydrogen
and the acid group $SO_4$. The ion $SO_4$, however, when it reaches
the anode, attacks the water, produces a molecule of $H_2SO_4$, and
liberates oxygen.

An interesting example of secondary action is furnished by
the common technical process of electroplating with silver from a
bath of potassium silver cyanide. The operation has been studied
by Hittorf among others, who holds that the cation is potassium,
and the anion the group $AgCy_2$. Each K ion, as it reaches the
cathode, precipitates silver by reacting with the solution in accordance with the equation

$$K + KAgCy_2 = 2KCy + Ag,$$

while the anion $AgCy_2$ dissolves an atom of silver from the anode,
and re-forms the complex cyanide $KAgCy_2$ by combining with the
$2KCy$ produced in the reaction described by the above equation.
If the anode consist of platinum, cyanogen gas is evolved thereat
from the anion $AgCy_2$, and the platinum becomes covered with the

insoluble silver cyanide AgCy, which soon stops the current. The coating of silver obtained by the process described above is coherent and homogeneous, while that deposited from a solution of silver nitrate, as the result of the primary action of the current, is crystalline and easily detached.

The corresponding cyanide process in the case of gold is now extensively used for the extraction of gold from its ores. The rock, containing small quantities of gold in a state of very fine division, is treated with potassium cyanide, and the solution of the double cyanide obtained in this way is electrolysed between steel anodes and lead cathodes. Prussian blue, which is again worked up into potassium cyanide, is formed on the anodes, and the gold is removed from the lead cathodes by cupellation.

Many organic compounds can be prepared synthetically by taking advantage of secondary actions at the electrodes, such as reduction by the cathodic hydrogen, or oxidation at the anode.

The injurious effects of polarization in primary batteries led to many attempts to overcome it. The methods in use in the common form of cell are well known, and have been described in § 36, Chapter v.

**79.** Davy had shown previously that there was no accumulation of electricity in any part of a voltaic circuit, and that a uniform flow or current existed throughout. Faraday set himself to examine the relation between the strength of this current and the amount of chemical decomposition. He first proved, by observations on the decomposition of acidulated water, that the amount of chemical action in each of several cells was the same when the cells were joined together and a current passed through them all in series, even if the sizes of the platinum plates were different in each. The volume of hydrogen was unchanged, even if electrodes of different materials—such as zinc or copper— were used. He then divided the current after it had passed through one cell into two parts, each of which passed through another cell before being reunited. The sum of the volumes of the gases evolved in these two cells was equal to the volume evolved in the first cell. The strength of the acid solution was then varied, so that it was different in the different cells in one series, but the

*Faraday's Laws.*

chemical action still remained the same in all. Thus the induction known as Faraday's first law was made :—

The amount of decomposition is proportional to the quantity of electricity which passes.

An apparatus for the decomposition of water can therefore be used to measure the total quantity of electricity which has passed round a circuit. Such instruments are termed voltameters.

The same law was then shown to be true for solutions of various metallic salts, and also for salts in a state of fusion—the weight of metal deposited being always the same for the same quantity of electricity. When the relative masses of the deposits of different substances by the same current were examined, a most important result appeared, which may be formulated as Faraday's second law :—

The mass of any substance liberated by a definite quantity of electricity is proportional to the chemical equivalent weight of the substance. In the case of elementary ions, this equivalent weight is the atomic weight divided by the valency, and, in the case of compound ions, it is the molecular weight divided by the valency.

It was then proved that the amount of zinc consumed in each cell of the battery was identical with that deposited by the same current in an electrolytic cell placed in the external circuit.

Faraday's work laid the foundations of the modern quantitative science of electrolysis. His results can be gathered into one statement, as follows :—

The quantity of a substance which separates at an electrode is proportional to the whole amount of electricity which passes, and to the chemical equivalent weight of the substance.

**80.** An accurate confirmation of Faraday's law for solutions of silver salts has been effected incidentally in the course of many experimental determinations of the electrochemical equivalent of silver. If the value obtained for the silver deposited by unit quantity of electricity be the same when the strength of current and the other conditions of the experiment are varied, the quantity of electricity and the mass of silver deposited must be proportional to each other. An exact knowledge of the electrochemical equivalent of silver is of great

Electro-
chemical
equivalents.

importance, since, given this constant, a silver voltameter can be used as a means of measuring accurately the total quantity of electricity, or the average current, which has passed through a circuit.

In order to determine the electrochemical equivalent, a constant current, which is measured simultaneously in absolute electromagnetic units (see § 62, Chapter VIII), is passed for a measured time through a solution of some silver salt. The most constant results are obtained when a neutral solution of the nitrate is used, containing about fifteen parts of salt to one hundred of water, and the current has an intensity of about one hundredth of an ampere to the square centimetre. The silver may be deposited on a platinum bowl used as cathode, the anode being a silver plate wrapped in filter paper to catch any particles disintegrated. The electrochemical equivalent is expressed as the number of grams of silver deposited by a current of one ampere in one second.

The mean result of the best determinations is about 0·001118 or 0·001119 gram per ampere-second. As we have stated (p. 186), the practical definition of the ampere assumes the value 0·001118.

The corresponding constant for other elements or compounds can be calculated from this number by dividing it by the chemical equivalent of silver, viz. 107·9, and multiplying by the chemical equivalent of the substance required. The value of hydrogen thus comes out $1·044 \times 10^{-5}$, its atomic weight being taken as 1·008, when oxygen is 16.

It will be noticed that the chemical constant involved is the equivalent, and not the atomic weight. Therefore, in the case of substances like iron, which form two series of salts, the amounts deposited will be different when solutions of the different salts are used. The two amounts will be in the proportion of the two chemical equivalents; if a current be sent through solutions of a ferric and a ferrous salt in series, the resultant weights will be as 56/3 : 56/2.

With no substance other than silver have such accurate experimental results been obtained, though many observations have been made on other bodies, solid and gaseous. In all cases, Faraday's laws have been found to be true within the limits of experimental error, and we may calculate electrochemical equivalents from the measured value for silver and the known chemical equivalents of the different ions. Kohlrausch and Holborn, in their book *Das Leitvermögen der Elektrolyte*, give a list of equivalent and electrochemical equivalent weights, the experimental value for silver being taken as 1·118 mg./amp.-sec.

*Equivalent weights A ($\frac{1}{2}$O = 8·00), and electrochemical equivalents E in mg./(amp.-sec.) of mono- and di-valent ions.*

| Cations | | | Anions | | |
|---|---|---|---|---|---|
| | *A* | *E* | | *A* | *E* |
| — | 1 | 0·01036 | Cl | 35·45 | 0·3673 |
| H | 1·008 | 0·01044 | Br | 79·96 | 0·8283 |
| K | 39·14 | 0·4055 | I | 126·86 | 1·3142 |
| Na | 23·05 | 0·2388 | Fl | 19·05 | 0·1973 |
| Li | 7·03 | 0·0728 | OH | 17·01 | 0·1762 |
| Ag | 107·92 | 1·118 | CN | 26·04 | 0·2698 |
| $NH_4$ | 18·07 | 0·1872 | $NO_3$ | 62·04 | 0·6427 |
| $\frac{1}{2}Ba$ | 68·70 | 0·7117 | $ClO_3$ | 83·45 | 0·8645 |
| $\frac{1}{2}Sr$ | 43·81 | 0·4539 | $BrO_3$ | 127·96 | 1·3256 |
| $\frac{1}{2}Ca$ | 20·02 | 0·2074 | $IO_3$ | 174·86 | 1·8115 |
| $\frac{1}{2}Mg$ | 12·17 | 0·1261 | $CHO_2$ | 45·01 | 0·4663 |
| $\frac{1}{2}Zn$ | 32·7 | 0·3388 | $C_2H_3O_2$ | 59·02 | 0·6114 |
| $\frac{1}{2}Cd$ | 56·05 | 0·5807 | $\frac{1}{2}O$ | 8·00 | 0·08288 |
| $\frac{1}{2}Cu$ | 31·8 | 0·3294 | $\frac{1}{2}S$ | 16·03 | 0·1661 |
| $\frac{1}{2}Fe$ | 28·01 | 0·2902 | $\frac{1}{2}SO_4$ | 48·03 | 0·4976 |
| $\frac{1}{2}Mn$ | 27·5 | 0·2849 | $\frac{1}{2}CrO_4$ | 58·07 | 0·6016 |
| $\frac{1}{2}Ni$ | 29·35 | 0·3041 | $\frac{1}{2}CO_3$ | 30·00 | 0·3108 |
| $\frac{1}{2}Pb$ | 103·46 | 1·0718 | $\frac{1}{2}C_2O_4$ | 44·00 | 0·4558 |
| $\frac{1}{2}Cr$ | 26·07 | 0·2704 | $\frac{1}{2}SiO_3$ | 38·20 | 0·3957 |

Solvents other than water, for example acetone, pyridine, liquefied hydrogen chloride, etc. have also been used. Faraday's laws hold good in such cases, and the electrochemical equivalents seem to be identical with those obtained when the solvent is water. Faraday's laws have also been demonstrated for fused salts, many of which are good electrolytes, with conductivities of the same order as those of aqueous solutions.

Again, in recent years it has been shown that, in certain cases, the discharge of electricity through gases is an electrolytic process accompanied by chemical decomposition. Here also the corresponding amount of electric transfer is accompanied by the same amount of chemical separation—the electrochemical equivalents are the same for gaseous electrolytes as for solutions.

In every case of electrolysis, Faraday's laws seem to apply, and the amount of a given substance liberated by a given transfer of electricity appears to be the same under all conditions. This result leads to an exact view as to the nature of the process. Since the amount of substance deposited is proportional to the quantity of electricity which passes, it follows that a definite charge of electricity is associated with a definite mass of the substance. We are thus led to look on the passage of an electric current through a solution as due to the carriage by moving parts of the electrolyte of opposite electric charges in opposite directions through the liquid. Each ion carries with it a fixed charge of electricity, positive or negative, which is given up to the electrode under the influence of an electromotive force above a certain limit. It is clear that, on this convective view of electrolysis, the conductivity of a solution must be proportional to the charge on each ion, to the number of ions, and to the velocity with which they move through the solution.

Whenever one gramme-atom or gramme-molecule of any monovalent ion is separated at an electrode, the same quantity of electricity passes round the circuit ; when the ion is divalent, the quantity is twice as great, and so on. All monovalent ions must therefore be associated with the same charge, all divalent ions with twice that charge, etc.

The quantity of electricity involved is easily calculated by considering an example. If a current of one ampere flows for one second, experiment shows that 0·001118 of a gramme of silver is liberated from the solution of one of its salts. Thus, when the equivalent weight in grammes is deposited, the quantity of electricity passing is 107·92/0·001118 or 96530 ampere-seconds or coulombs. The same result is of course true for the gramme-equivalent of any other substance, the gramme-equivalent being the gramme-molecule or gramme-atom divided by the valency. Whenever a gramme-equivalent of a substance is decomposed, therefore, 96530 coulombs of electricity pass round the circuit, and, as we shall prove later, this is the amount of charge actually transported through the electrolyte by one gramme-equivalent of any ion.

It is possible to calculate approximately the absolute electric charge carried by a single monovalent ion, since the number of molecules in a given volume of gas can be estimated by aid of the kinetic theory. As a mean value, at $0^\circ$ C. and normal atmospheric pressure, there are calculated to be about $6 \times 10^{19}$ molecules in one cubic centimetre of any gas.

As we have seen, one electromagnetic unit of electricity evolves $1\cdot044 \times 10^{-4}$ gramme of hydrogen, which at normal temperature and pressure fills a volume of $1\cdot16$ c.c., and therefore contains about $7 \times 10^{19}$ molecules or $1\cdot4 \times 10^{20}$ atoms, and yields the latter number of ions when dissolved as a hydrogen salt. Each ion is then associated with $7\cdot1 \times 10^{-21}$ electromagnetic units. The ratio between the units of electric quantity being $3 \times 10^{10}$, the ionic charge is about $2 \times 10^{-10}$ electrostatic units.

In a future chapter on the conduction of electricity through gases, we shall describe experiments by which the absolute charge on a gaseous ion may be estimated. The value obtained is about $3 \times 10^{-10}$ electrostatic units—identical within the limits of error with the absolute charge on an ion in liquid electrolytes.

Thus, the electric charge on a single monovalent ion seems to be a true natural unit, and the results we have summarized lead to an atomic theory of electricity. As von Helmholtz has said, "If we accept the hypothesis that the elementary substances are composed of atoms, we cannot avoid concluding that electricity also is divided into definite elementary portions, which behave like atoms of electricity."

**81.** The current through a metallic conductor is, to a very high degree of accuracy, proportional to the electro-motive force applied. As we have seen, this relation, known as Ohm's law, may be expressed in the form that $c = E/R$, where $R$ is a constant for any given conductor under fixed conditions, and is called its resistance. The law is verified if the resistance be shown to be independent of the current passing through it, and this method of proof has been used in the case of metallic conductors. The early experimenters, in the course of their investigations, made efforts to discover whether electrolytes also conformed to Ohm's law. It was known that, owing to the reverse force of polarization, no permanent current of moderate intensity could be maintained through an electrolyte unless the

*The conductivity of electrolytes.*

electromotive force exceeded a certain limit; but polarization occurs, primarily at any rate, at the electrodes, and it remained to see, when all reverse forces at the electrodes were eliminated, whether the flow of the current in the body of the liquid was in accordance with the law. Eventually Prof. F. Kohlrausch proved conclusively that solutions have a real resistance, which remains constant when measured with various currents and by different methods.

The current in a circuit containing an electrolytic cell can therefore be calculated by Ohm's law, if, from the total electromotive force of the circuit be subtracted the reverse electromotive force due to the polarization of the electrodes and to any changes produced by the current in the nature and concentration of different parts of the solution.

**82.** Owing to the difficulties introduced by polarization, the

Experimental methods.

resistance of an electrolyte cannot be measured by the continuous current methods adopted in the case of metallic conductors; it is necessary, in some way, to eliminate the effects of the reverse electromotive force at the electrodes. Many attempts were made before a satisfactory mode of experiment was devised, and developed into a convenient method.

If, instead of using continuous currents, we pass currents which alternate in direction through a solution, the products of the decomposition deposited on the electrodes by the passage of the current in one direction are removed by the reversed current which follows immediately. If the alternations be rapid enough, the quantity of substance deposited during each rush of current is very small. Moreover, when deposition first begins, and we are still far from the limiting value of the polarization, the reverse electromotive force is proportional to the surface density of the deposit. It may therefore be diminished again by increasing the area of the electrodes, and spreading the products of the decomposition over larger areas. The area of a platinum plate is increased enormously if we coat it with platinum black. In order to do this, a current is passed backwards and forwards between two electrodes through a solution of platinum chloride. By using currents which alternate rapidly, and electrodes coated with platinum black, the effects of polorization may be made insensible in our measurements of resistance.

With the usual form of Wheatstone's bridge, alternating currents would give no deflection of the galvanometer, and some modification is required. Kohlrausch used a telephone as indicator in place of the galvanometer; a rapidly alternating current gives a buzzing sound in the telephone, and the bridge is adjusted till this sound disappears, or is a minimum. The alternating currents may be obtained by the use of a small induction coil, or by passing the current of an alternating electric supply system through a suitable transformer to give an electromotive force of a few volts at most. The most usual form of apparatus is shown in Figure 99.

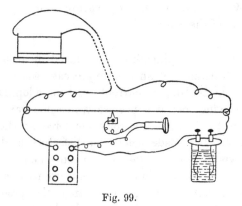

Fig. 99.

The metre bridge is adjusted till no sound is heard in the telephone, when the well-known relation between the resistances of the four arms of the bridge holds good.

The telephone is not a very pleasant instrument to use in this way, and a modification of the method is more rapid and also more accurate. The current from one or more dry cells is led to an ebonite drum, turned by a motor or a hand-wheel. On the drum are fixed brass strips with wire brushes touching them in such a manner that the current is reversed several times in each revolution. The wires from the drum are connected with an ordinary resistance box in the same way as the battery wires of the usual Wheatstone's bridge. A moving coil galvanometer is used as indicator, and on the other end of the drum there is another set of strips, arranged to reverse periodically the connexions of the galvanometer, so that any residual current which flows through it is direct and not alternating. These strips are rather narrower than the first set, and thus the galvanometer circuit is made just after the battery circuit is made, and broken just before the battery circuit is broken. The high moment of inertia of the galvanometer coil makes its period of swing very slow compared with the period of alternation of the current, and therefore the slight residual effects of polarization and other periodic disturbances are prevented from sensibly affecting the galvanometer. When the measured

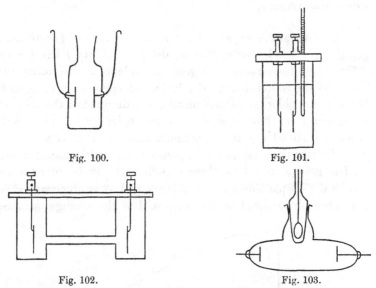

Fig. 100.                                   Fig. 101.

Fig. 102.                                   Fig. 103.

resistance is not altered by increasing the speed of the commutator, or changing the ratio of the arms of the bridge, the disturbing effects may be considered to be eliminated.

The form of vessel chosen to contain the electrolyte depends on the order of resistance to be measured. For dilute solutions the shapes of Figures 100 and 101 will be found convenient, while for more concentrated solutions those indicated in Figures 102 and 103 are suitable.

The absolute resistances of certain solutions have been determined by Kohlrausch by comparison with mercury, and, by using one of these solutions in any cell, the constant of that cell can be found once for all. From the observed resistance of any given solution in the cell, the resistance of a centimetre cube, or the specific resistance, can then be calculated. The reciprocal of this, or the conductivity, is a more generally useful constant; it is conveniently expressed in terms of a unit equal to the reciprocal of an ohm.

As the temperature coefficient of conductivity is large, usually about two per cent. per degree, it is necessary to place the resistance cell in a paraffin or water bath, and observe its temperature with some accuracy.

**83.** Kohlrausch expressed his results in terms of equivalent
Experimental conductivity, that is, the conductivity $k$ divided by
results. the number $n$ of gram-equivalents of electrolyte per litre. As the concentration of solutions of monovalent salts, such as potassium chloride, sodium nitrate, etc., diminishes, the value of $k/n$ approaches a limit, and, if the dilution be carried far enough in water distilled repeatedly, becomes constant, that is to say, at great dilution the conductivity is proportional to the concentration.

The general result of these experiments can be represented graphically by plotting $k/n$ as ordinates, and $n^{\frac{1}{3}}$ as abscissae; $n^{\frac{1}{3}}$ is a number proportional to the reciprocal of the average distance

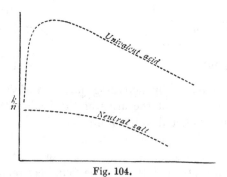

Fig. 104.

between the molecules, to which it seems likely that the equivalent conductivity will be related. The general forms of the curves for the neutral salt of a monovalent acid and for a caustic alkali or monovalent acid (like hydrochloric acid) are shown in Fig. 104.

The curve for the neutral salt comes to a limiting value, while that for the acid or alkali attains a maximum at a certain very small concentration, and falls again when the dilution is pushed to extreme limits. The meaning of this fall still remains to be elucidated. In some solvents the form of both these curves is entirely different.

The values of the equivalent conductivities of all neutral salts dissolved in water are, at great dilution, of the same order of magnitude, while those of acids at the maximum are about three times as great.

Passing to salts of divalent acids and other more complicated electrolytes, we find it impossible to reach such definite limiting values for the equivalent conductivity as are given by monovalent salts. Moreover, the influence of increasing concentration is more marked, the curves sloping at much larger angles. These changes in the phenomena are still greater when, as in copper sulphate, both metal and acid are divalent, and greatest of all in such substances as ammonia and acetic acid, which have very small conductivities when dissolved in water.

**84.** As we saw on p. 231, the experimental relations summarized in Faraday's laws indicate that electrolysis is to be considered as a process resembling convection, a constant stream of cations moving with the current, and a stream of anions in the opposite direction. The quantity of electricity thus conveyed will be proportional to the number of carriers and to the speed with which they travel.

*The migration of the ions and transport numbers.*

If we pass a current between copper plates through a solution of copper sulphate, the colour of the liquid in the neighbourhood of the anode becomes deeper, and in the neighbourhood of the cathode lighter in shade. This is well seen if the electrodes be arranged horizontally with the anode underneath. When the electrodes are of copper, the quantity of metal in solution remains constant, since it is dissolved from the anode as fast as it is deposited at the cathode, but, if we use platinum electrodes, the amount in solution continually becomes less. More salt is taken from the neighbourhood of the cathode than from the anode, and the colour of the solution becomes pale more rapidly near the cathode than near the anode.

This subject was first investigated systematically by Hittorf, who examined many solutions in a manner which enabled the liquid round the two electrodes to be analysed separately after the passage of the current.

We will assume at first that the ions are simple, or, at all events, that the opposite ions are associated with equal amounts of solvent or salt. If the opposite ions move with equal velocities, the result of the passage of the current will be that, while the composition of the middle portion of the solution remains unaltered, the products of the decomposition, which appear at the electrodes, are taken in equal proportions from the solution surrounding the anode, and from that round the cathode. If, however, one of the ions travels faster than the other, it will get away from the portion of the solution whence it comes more quickly than the other ion enters. When the electrodes are of non-dissolvable material, herefore, the concentration of the liquid in this region will fall faster than in that round the other electrode.

Let us assume that the cation drifts to the right with a velocity $u$, and the anion to the left with a velocity $v$. The velocity of the cation can be resolved into $\frac{1}{2}(u+v)$ and $\frac{1}{2}(u-v)$, and the velocity of the anion into $\frac{1}{2}(v+u)$ and $\frac{1}{2}(v-u)$. On pairing these components, we have a drift of the two ions right and left, each with a speed $\frac{1}{2}(u+v)$, involving no accumulation at the electrodes, and a uniform flow of the electrolyte itself without separation with a speed $\frac{1}{2}(u-v)$ to the right.

Thus at the cathode there is a gain of electrolyte equal to $\frac{1}{2}(u-v)$, and a loss, due to electrolytic separation, of $\frac{1}{2}(u+v)$: a total loss of $v$. At the anode there is a loss of $\frac{1}{2}(u+v)$ and a loss of $\frac{1}{2}(u+v)$, a total loss of $u$. The initial losses of electrolyte at the two electrodes, then, before diffusion sensibly affects the result, are in the same ratio as the velocities of the ions travelling away from them.

The process can be illustrated clearly by a method due to Hittorf. In Fig. 105 the black dots represent the one ion, and the white circles the other. If the black ions move to the left twice as fast as the white ions move to the right, the black ions will move over two of our spaces while the white ones move over one. Two of these steps are represented in the diagram. At the end of the process it will be found that six molecules have been decomposed, six black ions being liberated at the left and six white ions at the right. Looking at the combined molecules, however, we see that while five remain on the left side of the middle line, only three are still present on the right. The left-hand side, towards which the

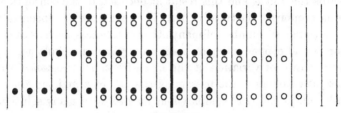

Fig. 105.

faster ions moved, has lost two combined molecules, while the right-hand side, towards which the slower ions travelled, has lost four—just twice as many. Thus we see that the ratio of the masses of salt lost by the two sides is the same as the ratio of the velocities of the ions leaving them. Therefore, on the assumption that no unsymmetrical complex ions are present, by analysing the contents of a solution after a current has passed, we can calculate the ratio of the velocities of its two ions. Hittorf called the phenomenon the "migration of the ions," and expressed his results in terms of a transport number, or migration constant, which gives the amount of salt taken from the neighbourhood of the cathode as a fraction of the whole amount that disappears. If there be no unsymmetrical complex ions, it also expresses the ratio of the velocity of the anion to the sum of the opposite ionic velocities.

The following table gives the transport numbers for some few salts; the concentration of the solutions being expressed as the number of gramme-equivalents per litre.

*Transport Numbers.*

| $n =$ | 0·01 | 0·1 | 1·0 | 2·0 |
|---|---|---|---|---|
| KCl | 0·506 | 0·508 | 0·514 | 0·515 |
| $AgNO_3$ | 0·528 | 0·528 | 0·501 | 0·476 |
| NaOH | — | 0·82 | 0·825 | — |
| HCl | — | 0·172 | 0·176 | 0·185 |
| $\frac{1}{2}CdI_2$ | 0·56 | 0·71 | 1·12 | 1·22 |
| $\frac{1}{2}CuSO_4$ | — | 0·632 | 0·696 | 0·720 |
| $\frac{1}{2}H_2SO_4$ | — | 0·191 | 0·174 | 0·168 |

The transport numbers of cadmium iodide, which, for solutions of more than half normal concentration, are greater than unity, show that the cathode vessel loses more salt than the whole solution does. It follows that some unaltered salt must travel through the solution towards the anode, and this result at once led to the conception of complex anions of the type $I(CdI_2)$. The changes with concentration in the transport number of many other substances, such as calcium chloride and copper sulphate, seem too great to be explained by a different rate of variation of the quasi-frictional resistance which the solution offers to the passage of the two ions, and suggest that similar unsymmetrical complex ions may exist in many solutions.

**85.** A further step was taken in the year 1879 by Kohlrausch, Mobility of who showed that a knowledge of the conductivity of the ions. a solution enables the sum of the opposite ionic velocities to be calculated. We have seen that we can represent the facts by considering the process of electrolysis to be a kind of convection, the ions moving through the solution and carrying their charges with them. Each monovalent ion may be supposed to carry a certain definite charge, which we can take to be the ultimate indivisible unit of electricity; each divalent ion carries twice that amount, and so on.

Let us consider, as an example, the case of an aqueous solution of potassium chloride, of which the concentration is $m$ gramme-equivalents per cubic centimetre. There will then be $m$ gramme-equivalents of potassium ions and the same number of chlorine ions in this volume. Let us suppose that on each gramme-equivalent of potassium there reside $+e$ units of electricity, and on each gramme-equivalent of chlorine ions $-e$ units. If $u$ denote the average velocity of the potassium ions, the positive charge carried per second across unit area normal to the flow is $meu$. Similarly, if $v$ be the average velocity of the chlorine ions, the negative charge carried in the opposite direction is $mev$. But positive electricity moving in one direction is equivalent to negative electricity moving in the other, so that the total current is $me(u+v)$.

Now let us consider the amounts of potassium and chlorine liberated at the electrodes by this current. At the cathode, if the chlorine ions were at rest, the excess of potassium ions would be the number arriving in one second, viz. $mu$. But, since the chlorine ions move also, a further separation occurs, and $mv$ potassium ions are left without partners. The total number of gramme-equivalents liberated therefore is $m(u+v)$. Now, by Faraday's law, the liberation of one gramme-equivalent of any ion involves the passage of a definite quantity $Q$ of electricity round the circuit. Thus, the total quantity passing in one second, that is the current, is $mQ(u+v)$. On comparing this result with the first expression for the same current, it follows that the charge, $e$, on one gramme-equivalent of either ion is equal to the quantity of electricity passing round the circuit when the gramme-equivalent is liberated.

We know that Ohm's law holds good for electrolytes, so that the current is also given by $-k\,dV/dx$, where $k$ denotes the conductivity of the solution, and $-dV/dx$ the potential gradient, *i.e.* the fall in potential per unit length along the lines of current flow.

Thus
$$me(u+v) = -k\frac{dV}{dx};$$

or
$$u+v = -\frac{k}{me}\cdot\frac{dV}{dx},$$

an equation in which everything may be expressed in centimetre-gramme-second units. By measuring the resistance $(= 1/K)$ in ohms (an ohm being $10^9$ C.G.S.), $e$ in coulombs $(10^{-1})$, and writing $n$ for the number of gramme-equivalents of solute per litre instead of expressing it per cubic centimetre, we get

$$u+v = -10^{-5}\frac{K}{ne}\cdot\frac{dV}{dx}.$$

Now $e$ is 96530 coulombs; so that for a potential gradient of 1 volt per centimetre ($10^8$ C.G.S. units), we have

$$u_1 + v_1 = 1{\cdot}036 \times 10^{-2} \times \frac{k}{n},$$

which gives the relative velocity (or the sum of the opposite velocities) of the two ions in centimetres per second under unit potential gradient. These numbers, $u_1$ and $v_1$, measure what we may call the mobilities of the two ions.

Since the transport numbers give us the ratio of the ionic velocities if no unsymmetrical complex ions are present, we can deduce the absolute values of $u_1$ and $v_1$ from this theory. Thus, for instance, the conductivity of a solution of potassium chloride, containing one-tenth of a gramme-equivalent per litre, is 0·01119 of a reciprocal ohm at 18° C. Therefore

$$u_1 + v_1 = 1·037 \times 10^{-2} \times 0·1119$$

$$= 0·001165 \text{ cm. per sec.}$$

Hittorf's experiments show us that the ratio of the velocity of the anion to that of the cation in this solution is ·51 : ·49. The absolute velocity of the chlorine ion under unit potential gradient is therefore 0·000595 cm. per sec., and that of the potassium ion 0·000570 cm. per sec. Similar calculations can be made for solutions of other concentrations and of other salts. An examination of the

| | | | | | |
|---|---|---|---|---|---|
| K ... | ... | $67 \times 10^{-5}$ | Cl ... | ... | $70 \times 10^{-5}$ |
| Na | ... | 45 ,, | I ... | ... | 70 ,, |
| Li ... | ... | 36 ,, | $NO_3$ | ... | 65 ,, |
| $NH_4$ | ... | 67 ,, | OH | ... | 184 ,, |
| H ... | ... | 323 ,, | $C_2H_3O_2$ ... | | 36 ,, |
| Ag | ... | 58 ,, | $C_3H_5O_2$ ... | | 33 ,, |

results shows that, in general, the velocities of the ions increase as the concentrations of the solutions diminish, and, at great dilution, the velocity of an ion in the solution in water of a simple binary salt is independent of the nature of the other ion present. From this result we may deduce the existence of specific ionic mobilities, the values of which are given in the table for different monovalent ions in centimetres per second per volt per centimetre.

**86.** Sir Oliver Lodge was the first to measure directly the

Experimental measurements of ionic velocity.

velocity of transport of an ion. In a horizontal glass tube connecting two vessels filled with dilute sulphuric acid he placed a solution of sodium chloride in solid agar-agar jelly. This solid solution was made alkaline with a trace of caustic soda to bring out the red colour of a little phenolphthalein added as indicator. A current was then passed from one vessel to the other along the tube. The hydrogen ions from the anode vessel of acid were thus carried along the tube, and decolorized the phenolphthalein as they travelled. By this method the velocity of the hydrogen ion through a jelly solution under a known potential gradient could be observed. The results of three experiments gave 0·0029, 0·0026, and 0·0024 cm. per sec. as the velocity of the hydrogen ion for a potential gradient of one volt per centimetre. Kohlrausch's number is 0·0032 for the dilution corresponding to maximum conductivity.

The velocities of other ions have been determined directly in another way by the present writer. Two solutions, having one ion in common, of equivalent concentrations, different densities, different colours, and nearly equal specific resistances, were placed one over the other in a vertical glass tube. An improvement on the original apparatus is shown in Fig. 106. The lighter solution is placed in the U tube, and the denser solution then run in below it. We may use, as an example, decinormal solutions of potassium carbonate and potassium bichromate. The colour of the latter is due to the presence of the bichromate group, $Cr_2O_7$. When a current is passed across the junction, the anions $CO_3$ and $Cr_2O_7$ travel in the direction opposite to that of the current, and their velocity may be determined by measuring the rate at which the colour boundary moves. The mobility of an ion is found to be very little less in a solid jelly than in an ordinary liquid solution. The velocities may be measured therefore by tracing the change in colour of an indicator, or the formation of a precipitate. Thus decinormal jelly solutions of barium chloride and sodium chloride, the latter containing a

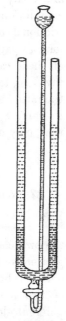

Fig. 106.

trace of sodium sulphate, may be placed in contact. Under the influence of an electromotive force, the barium ions move up the tube, and their presence is shown by the trace of insoluble barium sulphate formed. By keeping the conductivities of the two solutions nearly the same, discontinuity of potential gradient may be avoided, and the gradient may then be calculated from the area of cross-section of the tube, the conductivity of the solution, and the strength of the current as measured with a galvanometer.

These methods have been improved and extended by Orme-Masson and B. D. Steele. The general results confirm the results

16—2

of Kohlrausch's theory for simple binary salts. In solutions of more complicated electrolytes, the presence of unsymmetrical complex ions is suggested by other evidence, and the concordance of the direct measurements with Kohlrausch's theory is less exact.

As the concentration of dilute aqueous solutions increases, the equivalent conductivity falls, and the calculated ionic velocities with it. The direct measurements confirm this decrease of velocity and verify the calculation in the case of simple binary electrolytes. The diminution of conductivity may be due to an increase of the frictional resistance offered by the liquid to the passage of the ions, to part of the solute failing to be ionized, or to a combination of these causes. If we assume that the second cause alone is involved, the fraction $\alpha$ of the electrolyte ionized will be given by the ratio between the equivalent conductivity $\lambda$ of the solution, and its value $\lambda_\infty$ at infinite dilution. But it is important to remember that, only on the assumption named, may we write

$$\alpha = \frac{\lambda}{\lambda_\infty}.$$

**87.** When two metallic conductors are placed in an electrolyte, Voltaic a current will flow through a wire connecting them cells. provided that a difference of any kind exists between the two conductors in the nature either of the metals or of the portions of the electrolyte which surround them. A current can be obtained by the combination of two metals in the same electrolyte, of two metals in different electrolytes, of the same metal in different electrolytes, or of the same metal in solutions of the same electrolyte at different concentrations.

In order that the current should be maintained, and the electromotive force of the cell remain constant during action, it is necessary to insure that the changes in the cell, chemical or other, which produce the current, should neither destroy the difference between the electrodes, nor coat either electrode with a non-conducting layer through which the current cannot pass. As an example of a successful cell of fairly constant electromotive force, we may take that of Daniell, which consists of the electrical arrangement

zinc/zinc sulphate solution/copper sulphate solution/copper, the two solutions being usually separated by a pot of porous earthenware. When the zinc and the copper plates are connected through a wire, a current flows, the conventionally positive electricity passing from copper to zinc in the wire and from zinc to copper through the cell. Zinc dissolves, and zinc replaces an equivalent amount of copper in solution, copper being deposited simultaneously on the copper electrode. The internal rearrangements which accompany the production of a current do not cause any change in the original nature of the electrodes, and, as long as a moderate current flows, the only variation in the cell is the appearance of zinc sulphate on the copper side of the porous wall. While the supply of copper sulphate is maintained, copper, being more easily separated from its solution than zinc, is deposited alone at the cathode, and the cell remains constant. On the other hand, if no current be allowed to flow, slow processes of diffusion, unchecked by migration in the opposite direction, will cause copper to appear in the anode vessel, and finally to be deposited on the zinc. Little local galvanic cells are thus formed on the surface of the zinc, which then dissolves, even though the circuit of the main cell is not completed. Till this deposition occurs, the cell can be left on open circuit without waste, and no zinc will dissolve if it be chemically pure. If, however, commercial zinc, which contains iron, be used, local action is set up. This action can be prevented by amalgamating the zinc; probably because that process produces a uniform surface, iron being insoluble in mercury.

**88.** Considered thermodynamically, galvanic cells must be
Reversible          divided into reversible and non-reversible systems.
cells.              If the slow processes of diffusion be ignored, the Daniell cell already described may be taken as a type of a reversible cell. Let an electromotive force exactly equal to that of the cell be applied to it in the reverse direction. When the applied force is diminished by an infinitesimal amount, the cell produces a current in the usual direction, and the ordinary chemical changes occur. If the external electromotive force exceeds that of the cell by ever so little, a current flows in the opposite direction and all the former chemical changes are reversed, copper

dissolving from the copper plate, while zinc is deposited on the zinc plate. The cell, together with this balancing electromotive force, is thus a reversible system in true equilibrium, and the thermodynamical reasoning applicable to such systems can be used to examine its properties.

Cells from which gas is lost into the atmosphere, such as Volta's original couple, zinc/dilute acid/copper, and others in which irreversible processes of reduction occur, such as the Grove arrangement, zinc/dilute sulphuric acid/nitric acid/platinum, form essentially irreversible systems. Moreover, it does not follow that, because an accumulator can be used to give a current in the reverse direction to the charging current, it is, in the thermodynamic sense, a reversible cell. This is only the case when an electromotive force greater by an indefinitely small amount than the secondary electromotive force of the cell will reverse the current through it and the chemical action in it also. For this to be possible, it is necessary that the whole of the energy of the charging current should be put into available energy of chemical separation, which can all be regained when the cell is discharged.

Let us imagine that a reversible cell, balanced by an equal and opposite electromotive force, is put through a thermodynamic cycle of changes, after the manner of Carnot's engine. Let us draw an indicator diagram with ordinates denoting the electromotive force $E$ of the cell, and abscissae representing the electric transfer through the system. Let the system be placed in a chamber at a temperature $T$, and the balancing electromotive force be reduced infinitesimally, so that a quantity $e$ of electricity passes reversibly through the cell. The electrical work done is $Ee$. In the indicator diagram, Fig. 107, we travel along the isothermal line $AB$, and the cell absorbs a quantity $h$ of heat to keep its temperature constant.

Let us suppose that the system is thermally isolated, and that a further infinitesimal electric transfer occurs in the positive direction. The cell will either heat or cool; let us suppose that it cools to a temperature $T - \delta T$. If its electromotive force be continually balanced, the process will be reversible. Now let us place the system in an isothermal chamber at a temperature $T - \delta T$, and make the balancing electromotive force infinitesimally greater than that of the cell. An electric current flows in the negative direction, and the chemical changes of the cell are reversed. The indicating point in the diagram passes from $C$ to $D$, and a certain amount of electrical work is done on the cell. If, then, the system be thermally isolated, a further infinitesimal negative electric transfer will heat the cell reversibly to its original temperature $T$.

Now, if $\delta T$ is a very small temperature interval, the area of the figure $ABCDA$ in the indicator diagram—the area which gives the balance of electrical work done by the cell during the cycle—is independent of the particular shape of the ends, and is measured by the product of the length of the figure and its breadth. Whatever exact reversible operations are needed to take the cell from one isothermal line to the other, the electrical work is therefore

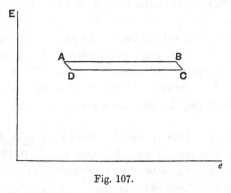

Fig. 107.

given by the product of the length and breadth of the figure. The length is $e$—the total electric transfer through the cell during either isothermal process—and the breadth is given by the difference in the electromotive forces at the two temperatures. This difference may be written as $\delta E$. Hence the useful electrical work gained from the cell during the cycle, or the area of the figure $ABCDA$, is $e\delta E$.

By the well-known expression for the efficiency of a reversible heat-engine, the ratio of the external work done to the energy $h$ absorbed from the source as heat during the hotter isothermal operation is equal to the ratio of the temperature range over which the engine works to the absolute temperatue of the source. Thus

$$\frac{e\delta E}{h} = \frac{\delta T}{T},$$

and, in the limit, when $\delta E$ and $\delta T$ are indefinitely small, we get
the relation

$$h = eT \frac{dE}{dT}.$$

This equation gives us the value of the reversible heat absorbed
by the cell during the isothermal production of a current; that is,
the heat which must be supplied at a temperature $T$ in order to
prevent the temperature changing when an amount $e$ of electricity
passes through the cell in the direction of its own electromotive
force. If the temperature coefficient of the electromotive force be
positive, that is, if the electromotive force of the cell increase with
increasing temperature, $h$ is positive also, and heat must be taken
into the cell to keep it isothermal. Left to itself, then, such a cell
will cool when giving a current. On the other hand, if $dE/dT$
be negative, the cell will be heated by an electric transfer. Both
these effects are, of course, independent of the non-reversible
Joule effect, by which an amount of heat $C^2Rt$ is developed by
the passage of a current. If the current be made very small,
the Joule effect may be neglected compared with the reversible
effect.

The analogy of this reversible effect with that discovered by
Peltier at the junction of two different metals (§ 49) is manifest,
and suggests that the site of the present phenomenon is to be
sought at the interfaces between the electrodes of the cell and the
liquids contained therein. This conclusion has been verified experimentally by Jahn and by Gill, who find that the local thermal
effects are in accordance with theory.

When a reversible voltaic cell is giving a current, the source of
the energy is clearly the chemical action. If the same amount of
chemical action were to take place in a calorimeter, without the performance of external work, the system might be brought to the same
final state, with the same change of internal energy. In this case,
all the energy that is converted goes into heat, and the calorimetric
heat $H$ evolved is a measure of the change in internal energy.
When used to drive a current, the same decrease in internal energy
must supply the electrical energy $Ee$, and also any reversible heat
evolved at the junction in the circuit. "Heat evolved" means
heat leaving the system, and thermodynamically is to be reckoned

negative. But, as we have seen, the reversible heat $h$ is equal to $eT dE/dT$. Hence we have

$$H = Ee - eT\frac{dE}{dT}.$$

If $\lambda$ denote the calorimetric heat evolved by an amount of chemical action corresponding with unit electric transfer, $H$ may be replaced by $\lambda e$, and we obtain the relation

$$\lambda = E - T\frac{dE}{dT},$$

or

$$E = \lambda + T\frac{dE}{dT},$$

an expression for the electromotive force of a voltaic cell, which was obtained by von Helmholtz, and, in a different way, by Willard Gibbs.

This equation is a particular example of the general equation connecting the change in the internal energy of a system with what is known as the free or available energy. The available energy $A$ is the amount of external work obtainable by an infinitesimal, isothermal, reversible change in the system, and, in the case considered, is $Ee$, or, for unit electric transfer, $E$, the electromotive force. The corresponding change $I$ in the internal energy is $\lambda$, and we may write our equation in the general form

$$A = I + T\frac{dA}{dT},$$

which is applicable to any reversible physical or chemical system.

Returning to the consideration of the equation for the electromotive force, it will be noticed that, if $dE/dT$ be zero, the electromotive force is measured by the heat of reaction per unit electrochemical change. The earliest formulation of the subject, due to Lord Kelvin, assumed that this relation was true in all cases; and, calculated in this way, the electromotive force of Daniell's cell, which happens to possess a very small temperature coefficient, agreed with observation.

When one gramme of zinc is dissolved in sulphuric acid, 1670 thermal units are evolved. For the electrochemical unit, or 0·003388 gramme, the thermal evolution is 5·66 calories.

Similarly, the calorimetric heat corresponding to the electro-chemical unit of copper is 3·00 calories. Hence, the thermal equivalent of the unit of electrochemical change in Daniell's cell is 2·66 calories. The dynamical equivalent of the calorie is $4·18 \times 10^7$ ergs, and the electromotive force should be $1·112 \times 10^8$ C.G.S. units or 1·112 volts—a close agreement with the experimental result of about 1·08 volts.

For cells in which the electromotive force changes with temperature, the accuracy of the equation of Helmholtz and Gibbs has also been confirmed experimentally.

**89.** As stated above, an electromotive force is produced when-

Concentration cells.

ever there is a difference of any kind at two elec-trodes immersed in electrolytes. In ordinary cells the difference is secured by using two dissimilar metals, but an electro-motive force exists if two plates of the same metal are placed in solutions of different substances, or of the same substance at different concentrations. Another method is to use in the same solution electrodes of different concentration. Such electrodes can be constructed by taking hydrogen in contact with plantinized plantinum, and making the pressure different at the two ends. In all such cells the electrical energy is not obtained from chemical changes, but from the energy of expansion of substances from greater to smaller concentrations.

Let us take as an example of a concentration cell the arrange-ment

silver/dilute silver nitrate/concentrated silver nitrate/silver.

Here metal dissolves in the more dilute solution, and is deposited from the more concentrated solution, and this process will continue, since it involves a decrease of available energy, till the concentra-tions are equalized.

When one electrochemical unit of electricity passes, one gramme-equivalent of silver dissolves at the anode, and an equal quantity is deposited at the cathode. In this manner, the anode vessel must gain one gramme-equivalent of salt and the cathode vessel lose the same amount. Now let us consider the motion of the ions through the solution. The current, which is exclusively carried by silver ions at the electrodes, is shared between silver ions and $NO_3$

ions in the body of the liquid. If the ionic velocities were the same, therefore, half a gramme-equivalent of each ion would pass across the surface of contact of the solutions. In the general case, when the transport ratio of the anion is $r$, and that of the cation $1 - r$, the anode vessel will, on the whole, gain $1 - (1 - r)$ or $r$ gramme-equivalents of silver and therefore of salt, while the cathode vessel must lose an equal amount, the difference between this case and that considered on p. 237 consisting in the fact that now we have a dissolvable anode.

In order to apply thermodynamical reasoning to this case, and so obtain an expression for the electromotive force, we must, in some way, complete the cycle of operations, and undo the changes which have occurred in the cell; we must re-concentrate the solution which has lost salt, and re-dilute the one which has gained salt. This process may be performed in one of two ways. We may evaporate solvent from the one solution, and re-condense it again on to the other, or we may effect the same result by what is called an osmotic process.

As we shall see later (§ 96), when confined in a cell permeable to solvent but not to solution, a dissolved substance sets up, by the entrance of solvent, a certain pressure called the osmotic pressure. Both theoretically, as a consequence of thermodynamics, and practically, by experiment in cases such as sugar solutions, the osmotic pressure has been shown to be equivalent to the pressure exerted by a gas which, per unit volume, contains the same number of molecules as the solution contains molecules of solute.

Let us return to the consideration of our concentration cell. By the electrical process, which involves unit electric transfer through the cell, $r$ equivalents of salt have been transferred from the concentrated to the dilute solution. It remains to describe some process by means of which we may imagine the $r$ equivalents of salt returned from the dilute to the concentrated solution.

Let us place the more dilute solution, which has received additional salt by reason of the electric transfer, in an engine-cylinder, of which the bottom is pervious to the solvent but not to the salt, and is backed by a large volume of the pure solvent. Let the pressure on the piston be that of equilibrium. Allow this pressure to fall by an infinitesimal amount, so that solvent enters the solution till the concentration is again exactly as it was before the electric transfer.  The change in concentration is very small if a large volume of solution be present, so that the process practically occurs at constant pressure and the work gained is $P_1 v_1$, where $v_1$ denotes the change in volume, and $P_1$ the constant osmotic pressure. Now separate bodily from the solution that volume of it which contains the amount of salt transferred by the current, and reversibly compress this quantity in an osmotic cylinder till its osmotic pressure rises to $P_2$, that of the more concentrated solution of the cell.  The work done by the osmotic forces is $-\int_{P_1}^{P_2} P\,dv$. Finally place this liquid in contact with the stronger cell solution, connect it through a semi-permeable wall with the reservoir of the pure solvent, and squeeze out solvent till the solution regains its initial volume by the expenditure of work equal to $P_2 v_2$.  The thermodynamic cycle is then complete.

Since both the electrical and the osmotic processes of this cycle can be made reversible and isothermal, the balance of external work must vanish.  Denoting the electromotive force by $E$, and considering the electric transfer $e$, we may write

$$Ee + P_1 v_1 - \int_{P_1}^{P_2} P\,dv - P_2 v_2 = 0.$$

If we restrict ourselves to dilute solutions, for which the gaseous laws hold good, $P_1 v_1$ must be equal to $P_2 v_2$ and we have

$$Ee = \int_{P_1}^{P_2} P\,dv.$$

Now, as we shall see later, the osmotic pressure of dilute solutions of electrolytes has a value such that, on the gaseous laws, we must suppose that each ion of the salt produces its own pressure effect.  If $n$ be the number of ions given by one molecule of the salt, the pressure will, in very dilute solutions, be greater than that of an equivalent solution of a non-electrolyte in the ratio of $n$ to 1.

Again, $e$ is the electric transfer needed to liberate one gramme-atom of two opposite monovalent ions at the electrodes, and therefore to decompose one gramme-molecule of a monovalent salt.  If the salt does not yield two opposite monovalent ions, let $y$ be the total valency of the anions or of the cations obtained from one molecule; for instance, $y$ will be 2, whether the cations be two monovalent ions such as the two $H'+$ ions of a molecule of sulphuric acid, or one divalent ion such as the $Cu''+$ of copper sulphate. The total electric transfer corresponding with the decomposition of one gramme-molecule of salt, and the liberation of one gramme-atom of each of the ions, is $ey$.

For dilute solutions, the usual gaseous laws apply, and, with non-electrolytes, we may write the equation

$$Pv = RT,$$

where $v$ is the volume occupied by one gramme-molecule of the solute, $T$ is the absolute temperature, and $R$ is a constant. Keeping the same meaning for $T$, $R$, and $v$, in the case of dilute electrolytes, the pressure is greater in the ratio of $n$ to 1, where $n$ is the number of ions given by one molecule of the electrolyte. We then have

$$P = \frac{nRT}{v}.$$

In the concentration cell we have described, instead of one gramme-molecule, we have to deal with a transfer of $r$ gramme-molecules, and we must write

$$P = \frac{nrRT}{v}.$$

The equation between the electrical and the osmotic work, namely,

$$Eey = \int_{P_1}^{P_2} P\,dv,$$

now becomes
$$E = \frac{nrRT}{ey} \int_{v_1}^{v_2} \frac{1}{v}\,dv$$

$$= \frac{nrRT}{ey} \log_e \frac{v_2}{v_1}$$

$$= \frac{nrRT}{ey} \log_e \frac{P_1}{P_2}.$$

In dilute solutions, the osmotic pressures are proportional to the concentrations $C_1$ and $C_2$. Hence we obtain, as an expression for the electromotive force of our concentration cell,

$$E = \frac{nrRT}{ey} \log_e \frac{C_1}{C_2}.$$

The numerical value of this expression may be calculated. For decinormal and centinormal solutions of silver nitrate, the transport number $r$ is the same, and is equal to $0\cdot528$. In a cell containing these liquids, at a temperature of $18°$ C. or 291 absolute,

$$E = \frac{2 \times 0\cdot528 \times 8\cdot28 \times 10^7 \times 291}{9653 \times 1} \times 2\cdot303 \times \log_{10} 10,$$

since $R$, the gas constant per gramme-molecule, has the value $8\cdot28 \times 10^7$. Hence

$$E = 0\cdot060 \times 10^8 \text{ c.g.s. units}$$

$$= 0\cdot060 \text{ volt.}$$

Nernst measured the electromotive force of this cell experimentally, and found the value

$$E = 0\cdot055 \text{ volt.}$$

This number is in remarkable agreement with the result deduced theoretically. It confirms the general accuracy of the thermodynamic treatment of the subject given above, and shows that the assumption that solutions may be treated as "dilute" is justified for this purpose at the concentrations chosen.

The concentration cell which we have taken hitherto as an example, namely,

silver / dilute silver nitrate / concentrated silver nitrate / silver,

depends on the migration number for the anion. Cells have also been devised in which the electromotive force depends on the migration number for the cation, and other cells in which the effects of migration are altogether eliminated. An equally good agreement between theory and observation is found in these cases also.

The logarithmic formulae for all these concentration cells indicate that theoretically their electromotive force can be increased to any extent by diminishing without limit the concentration of the more dilute solution; $\log P_2/P_1$ then becomes very great. This condition can to some extent be realized in a manner that throws light on the general theory of the subject.

Let us consider the arrangement

Ag / AgCl with normal KCl / KNO₃ / decinormal AgNO₃ / Ag.

The concentration of silver chloride is very small in saturated aqueous solution; from the electric conductivity it has been estimated as 0·0000117 normal. It is still further reduced by the presence of the large excess of chlorine ions of the potassium chloride.

The electromotive force of this cell has been calculated as 0·52 volt. This number was confirmed experimentally by Ostwald, who also examined other cells with similar electrodes giving high electromotive forces.

Hittorf has shown that the effect of a cyanide round a copper electrode is to combine with the copper ions. The concentration of the simple copper ions is then so much diminished that the copper plate becomes an anode with regard to zinc. Thus the cell

$$Cu / KCN / K_2SO_4 / ZnSO_4 / Zn$$

furnishes a current which carries copper into solution and deposits zinc. In a similar way silver could be made to act as anode in presence of cadmium.

**90.** The principles used in developing the theory of concentration cells show that the electromotive force of a Daniell's cell will be raised by diminishing the concentration of the zinc sulphate solution, into which zinc dissolves, or by increasing the concentration of the copper sulphate solution, from which copper is precipitated. Compared with the total electromotive force, due chiefly to the chemical energy, these changes are of course small.

*Chemical cells.*

Similar ideas have been applied to the chemical changes themselves by Nernst and others, though in this case the basis of the investigation is more speculative. When a metal is placed in contact with the solution of one of its salts, and a current is passed across the junction and metal dissolved, changes occur in the chemical, osmotic, and electrical energies of the system. As the osmotic pressure of the solution rises, the tendency of the metal to dissolve as electrolytic ions becomes less, and it is suggested that eventually, at a certain pressure, no further tendency to dissolve would exist. Above this pressure the metal would tend to come out of solution and be deposited. This critical pressure bears no relation to the limit set to the osmotic pressure of a solution by reason of the finite solubility of the salt. With some metals it may be much too high to be reached, with others it may be too low. If the concentration of the solution giving the critical pressure could be obtained, so that there would be no tendency for ions of the metal to enter or leave the liquid, it is fair to conclude that the metal and solution would be electrically neutral to each other, and that no difference of potential would exist between them. This critical osmotic pressure has been called the electrolytic solution pressure of the metal in the given solution. Nernst identifies it with the osmotic pressure of the ions of the metal in the substance of the metal itself. Such an idea is perhaps suggested by the osmotic pressure of certain metals when dissolved in mercury to form amalgams.

**91.** Any reversible cell can theoretically be employed as an accumulator; though, in practice, conditions of general convenience are more sought after than thermodynamic reversibility.

Secondary cells or accumulators.

The accumulator commonly used can be made by placing two lead plates in dilute sulphuric acid, and passing a current between them. Hydrogen is evolved at the cathode, while the anode becomes covered with a layer of insoluble lead peroxide.

The cell is now in a condition to give a current in the reverse direction, during which process lead sulphate is formed at both electrodes until these electrodes become identical in constitution. In a second charging, the lead sulphate at the cathode is reduced to spongy lead, while at the anode it gives peroxide as before.

The mass of spongy material at the electrodes is increased by continual charging and discharging, a process which adds to the effective capacity of the cell; and the whole preliminary process of forming the cell can be hastened if the plates receive in the first place a coating of red lead, $Pb_2O_3$.

The main chemical action of a fully formed accumulator seems to be in accordance with the equation

$$PbO_2 + Pb + 2H_2SO_4 \rightleftarrows 2PbSO_4 + 2H_2O,$$

which, read from left to right, describes the discharge, and, from right to left, the charging of the cell. Although ozone, hydrogen peroxide, persulphuric acid, and traces of lead persulphate have been detected, it seems likely that the above equation represents the chief part of the changes. The concentration of the acid solution is an important factor in determining the electromotive force, which increases with increasing concentration, since part of the available energy of the reaction is due to the dilution of the residual acid by the water formed.

It is found in practice that the effective electromotive force of a secondary battery is less than that required to charge it; the energy efficiency of a lead accumulator is from 75 to 85 per cent., although from 94 to 97 per cent. of the current used in charging it can be regained. This drop in the electromotive force has led to the belief that thermodynamically the cell is only partly reversible. Dolezalek, however, has attributed the discrepancy to mechanical hindrances, which prevent the equalization of acid concentration in the neighbourhood of the electrodes, rather than to any essential irreversible chemical action.

On the provisional hypothesis that the system may be treated as reversible, the Gibbs-Helmholtz equation

$$E = \lambda + T\frac{dE}{dT}$$

has been applied. The discharge reaction for dilute solutions gives a calorimetric heat-evolution of 87,000 calories, which, on the assumption that the energy is all available, is equivalent to an electromotive force of about 1·88 volts. This number agrees with that observed for cells filled with weak acid, and indicates that the temperature coefficient is very small, a result which has been confirmed experimentally.

**92.** The source of the energy of a galvanic cell is certainly the
Volta's chemical action, a correction being applied for any
contact effect. reversible heat which the cell absorbs from or gives up to its surroundings. The exact seat of the difference of potential, however, remained undetermined for a century, and proved a fruitful subject of discussion. Volta located it at the junction of the unlike metals; while Faraday's work, which showed the regular and funda-

mental part played by the chemical processes, seemed to indicate the surfaces at which the metals were in contact with the liquids. These two views of the nature of the phenomena have continued till recent years, though it seems certain that a considerable difference of potential exists at the surface of separation between metals and electrolytes or dielectrics such as air. The facts to be explained, besides those of the galvanic cell, are as follows.

Dry zinc and copper brought into contact with each other in dry air become oppositely charged, and, if their substances are arranged parallel and very close to each other, so as to form a condenser of large capacity, these charges may be considerable. They can be exhibited by separating the plates; the capacity is then diminished, and the difference of potential is thereby increased, so that it can be indicated by an electroscope or measured by an electrometer. The natural potential-difference produced by the contact of zinc and copper is about three-quarters of a volt.

Many experiments have been made on this subject. Ayrton and Perry have examined many metals, obtaining among others the following potential-differences in volts :

| | |
|---|---|
| Zinc | 0·210 |
| Lead | 0·069 |
| Tin | 0·313 |
| Iron | 0·146 |
| Copper | |
| Platinum | 0·238 |
| Carbon | 0·113 |

By the summation of potential-differences, a principle experimentally established by Volta, we can find the contact effect between any two metals in this list by adding together the values for all the pairs of intervening metals.

The theory of the cell given above suggests that the chief potential-difference is to be sought at the liquid-metal surface; but it is clear that, before any such interpretation can be accepted, it must be reconciled with the phenomena of contact electricity just described. On the analogy of the cell, the most natural explanation is that the contact potential-difference is due to the action of the oxygen of the air. It is not necessary to imagine actual oxidation; a sufficient cause might possibly be found in some slight modification of the film of condensed gas, which seems to exist on all solid surfaces, and to be so difficult to remove.

When, instead of the insulator air, the plates are surrounded by an electrolytic conductor, the slope of potential is accompanied by a current through the solution. At the contacts of the liquid with the metals, the natural potential difference is constantly tending to be set up again by the chemical affinity; thus a constant current is maintained, and zinc is actually dissolved.

The probability that the contact effect depends on the chemical action or affinity of the surrounding medium will be much increased if it can be shown that the magnitude of the effect depends on the nature of the medium. No change is produced by working in vessels at high exhaustion, or by placing the metals in an atmosphere of hydrogen. But the film of gas which clings to a solid is exceedingly difficult to remove, and it now seems likely that its persistence explains these negative results. From a research by Spiers, in which extraordinary precautions were taken to remove the film of gas, it is clear that the difficulties of getting rid of it have been greatly undervalued, and that, when it is really disturbed, large changes in the magnitude of the contact force are produced. Brown has shown that, if the copper and zinc plates are intensely heated in oil—a process which would remove the surface films—Volta's effect is much diminished. In such a case a few experiments that yield a positive result, and indicate a probable reason for the negative results of others, seem to carry great weight.

The phenomena of thermo-electricity have an intimate connexion with the subject now under consideration.

Let us consider a complete circuit of two metals, $T_1$ and $T_2$ being the temperatures of the junctions. Let $\Pi_1$ and $\Pi_2$ be the heat evolved by unit electric transfer at the junctions at the temperatures $T_1$ and $T_2$ respectively; $\Pi_1$ and $\Pi_2$ measure the Peltier effects. Let $\sigma$ and $\sigma'$ denote the coefficients of the Thomson effect in the two metals. In Chapter VI we showed that, considering unit electric transfer round the circuit, the energy principle leads to the result

$$E = \Pi_1 - \Pi_2 + \int_{T_1}^{T_2} (\sigma - \sigma')\, dT,$$

where $E$ is the total electromotive force round the circuit, while the consideration of the entropy of the system gives us

$$\Pi = T \frac{dE}{dT}$$

for the Peltier effect at one junction.

Thus, while the Peltier effects depend on the temperature coefficient of the total electromotive force of the circuit, in the equation for $E$ the Peltier effect appears as a local electromotive force at the junction. Each electric unit, as it passes across, introduces an energy effect $e\Pi$, which involves a reversible evolution or absorption of heat. The Peltier effects have been determined experimentally, and their electrical equivalents, which measure the contact potential-differences at the metallic junctions, are calculated as a few millivolts only, values much too small to explain by themselves the observed Volta effects, without taking account of similar effects at the surfaces of the surrounding dielectric.

**93.** Many attempts have been made to determine experimentally the single potential-differences at the individual junctions in a circuit containing electrolytes as well as metals. In a galvanic cell, for example, there must be at least two such junctions, and the problem is to separate their effects and measure the step of potential at each. The measured electromotive force gives the sum of all the single potential-differences but the impossibility of connecting directly the electrolyte with an electrometer, without introducing another metallic junction, throws difficulties in the way of observing them individually. If a method could be devised capable of application to one such junction, the combination of that junction with any other would enable the value for the other to be calculated from the total electromotive force as observed in the usual manner.

*Single potential-differences at the junctions of metals with electrolytes.*

It has usually been thought that the single potential-difference at the common boundary of mercury and an electrolyte has been determined satisfactorily by experiments on capillary electrometers and by others in which mercury was allowed to drop into a solution from a fine glass nozzle.

Von Helmholtz pointed out that a potential-difference at the junction of mercury with an electrolyte might be explained by the supposition of a double layer of electricity over the surface, the two opposing faces being oppositely charged—on the side of the electrolyte by the congregation of ions. Such a system would take time to reach its final state, owing to the slow movement of the ions, and he concluded that, if mercury were allowed to drop rapidly from an orifice beneath the surface of a liquid electrolyte, the double layer would not be established, and the stock of falling mercury would be brought to the same potential as the electrolyte. The apparatus therefore might be used in connexion with an electrometer, the other quadrants being joined to a quantity of mercury at rest in the same electrolyte. A difference of potential of about 0·8 or 0·9 volt is obtained between the dropping mercury and the mercury at rest in dilute sulphuric acid, but there seems some doubt whether or not the dropping electrode is really at the potential of the electrolyte, and the trustworthiness of the method cannot yet be regarded as established.

Another method by which the measurement of single potential differences has been attempted depends on electrocapillary phenomena. The surface tension of the area of contact between the mercury and a solution is affected by its electrical state. If the surface be increased, an electric transfer is produced, and, conversely, if an external electromotive force be applied across the junction, the area tends to change, owing to an alteration in the effective surface tension. These phenomena have been applied by Lippmann to the construction of capillary electrometers, of which several forms are in frequent use. In one variety, a vertical glass tube is drawn to a very fine capillary. The tube is partially filled with mercury, and the lower portion immersed in an electrolyte, usually dilute sulphuric acid, in which is placed another quantity of mercury. The capillary

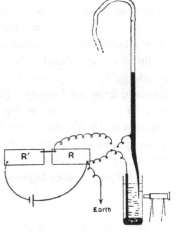

Fig. 108.

forces tend to raise the mercury surface in the little tube, and are balanced by the pressure of the long column. When the mercury in the vertical tube and the mercury below the electrolyte are connected with two conductors at different potentials, such as the opposite poles of a galvanic cell, a change is produced in the level of the surface of contact in the capillary tube. A microscope with a micrometer eyepiece may be arranged to view the capillary, and, for small differences, the change in level is found to be proportional to the applied difference of potential. While the applied electromotive force is a small fraction of a volt, the electrometer behaves as a condenser of good insulation, retaining its charge for several hours when disconnected from the cell. On the other hand, when the applied voltage is greater, electrolysis at the surface seems to occur, and the charges leak away.

An explanation of these phenomena, based on Lippmann's observations, has been given by von Helmholtz, on the assumption that no electrolysis occurs. Any natural potential difference between two bodies implies an electrification over the boundary, in such a manner that an electric condenser of minute thickness is formed, with its parallel faces oppositely charged. This electric double layer will produce an electrostatic surface energy $\epsilon$, the value of which is $\frac{1}{2}CAV^2$, where $C$ is the capacity of the double layer per unit surface, $A$ the area of contact, and $V$ the potential difference across the layer. Now if $V$ be kept constant, and the area be increased, we have for $d\epsilon/dA$ the value $\frac{1}{2}CV^2$. This increase in available energy is obtained from the chemical energy which maintains the natural potential difference, just as the increase of energy in the quadrant electrometer described on page 47 is obtained from the battery. In that case, the needle tends to move so as to cover more of the oppositely charged quadrants; and here the area of the double layer, that is, the area of contact between mercury and electrolyte, tends to increase. This effect therefore acts in a sense opposite to that of the ordinary surface tension $S_0$. The resultant surface tension $S$ will be

$$S = S_0 - \tfrac{1}{2}CV^2.$$

On the assumption that the only effect of the potential difference is to produce such an electrostatic effect, $S_0$ will be independent of

$V$, and the total observed surface tension will reach a maximum when $V$ is zero.

Lippmann and others have found that, in agreement with the equation given above, the curves drawn between the external electromotive force and the reading of a capillary electrometer rise to a maximum and then fall again, their form being roughly parabolic; with dilute sulphuric acid, the maximum of the curve is reached when about one volt is applied.

By using liquid amalgams, the same electrolyte can be compared with mercury and what is effectively a different metal. Rothmund and others have compared the difference of the values thus measured with the electromotive force of the cell

amalgam / electrolyte / mercury,

finding concordant results.

The results of these experiments on the whole favour the view that the electromotive force of a cell is approximately the sum of the single potential differences at its electrodes as determined by the capillary electrometer. On the other hand, there is evidence to show that, in general, the result of an applied electromotive force on the surface tension is not merely the electrostatic effect contemplated by the Helmholtz theory, but that it depends also on the chemical nature of the electrolyte. Whichever side of the electric double layer is positive, its effect is opposite to that of the natural surface energy, but there is no reason to suppose that the effects of anions and cations on the electrolyte side of the double layer will be identical. The total electrocapillary curve therefore probably consists of parts of two parabolas which meet at a point, and a slight want of symmetry observed in the ascending and descending branches of the experimental curves is thus explained. Van Laar, who has considered this effect in detail, concludes that single potential differences cannot be determined accurately by means of the capillary electrometer.

It is evident from what has been said that there is some doubt whether the experiments on dropping electrodes and on capillary electrometers really enable us to calculate even approximately the natural potential difference which is involved in the electromotive force of a galvanic cell containing a mercury-electrolyte surface.

Nevertheless, since many useful determinations of other single potential differences, which, at all events, are relatively exact, rest on such measurements, in the present condition of the subject we may provisionally accept the value of about $+0.92$ volt as the potential difference between mercury and dilute sulphuric acid, the mercury being positive to the acid. The step of potential as thus measured is in the opposite direction to that which occurs at the surface of a zinc plate. Results are obtained for other metal-electrolyte surfaces by subtracting this number, or another similarly estimated for mercury in contact with some other electrolyte, from the total electromotive force of galvanic cells arranged in the manner

metal / electrolyte / mercury.

Such indirect determinations will contain as a constant error any deviation of the primary measurement from the true value, but, as relative numbers, serving to compare the metals among themselves, they will retain their importance.

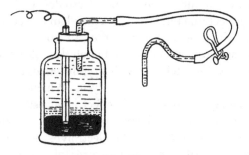

Fig. 109.

In making such experiments, it is usual to employ what is known as a normal electrode, consisting of a quantity of pure mercury covered by a layer of mercurous chloride and a solution of potassium chloride of normal concentration, that is, a solution containing one gramme-equivalent per litre. An indiarubber tube ending in a glass tube leads from the solution and is filled with it (Fig. 109). Contact can thus be made between the potassium chloride and any other liquid. This electrode, as measured by Lippmann's method, gives a potential difference of $0.56$ volt, the mercury tending to come out of solution and be deposited as metal.

The chlorides can of course be replaced by other substances when their potential with respect to mercury is known. Thus a soluble sulphate, with mercurous sulphate as depolarizer, has been used. Assuming that we may neglect the small effects at the junction of the metals, and at the surfaces of contact of unlike solutions, if such surfaces are present, the measured electromotive force of the combination metal / electrolyte / normal electrode enables the potential difference at the surface metal / electrolyte to be calculated by subtraction.

In this manner, Neumann measured the single potential differences for many metals in contact with either normal or saturated solutions of their salts. The following are some of the most important results.

| Metal | Sulphate | Chloride | Nitrate |
|---|---|---|---|
| Zinc ... | − 0·524 | − 0·503 | − 0·473 |
| Iron ... | − 0·093 | − 0·087 | — |
| Lead ... | — | + 0·095 | + 0·115 |
| Copper ... | + 0·515 | — | + 0·615 |
| Silver ... | + 0·974 | — | + 1·055 |
| Mercury ... | + 0·980 | — | + 1·028 |

In this table positive signs have been assigned to those metals which show a positive potential relatively to the liquids surrounding them. Assuming the accuracy of these results as absolute numbers, it follows that such metals tend to come out of solution, and the natural potential difference at their surfaces helps to drive a current in the direction to effect the deposition. Spontaneous separation of these metals, or solution of negative metals, however, will only occur if means be available for the simultaneous addition of opposite ions, or the removal of an equivalent quantity of similar ions.

For slightly oxidizable metals, such as silver, it will be seen that the method leads to numbers which have a sign opposite to that assigned to the values for very oxidizable substances such as zinc.

If we regard the potential differences as due to the potential energy of possible chemical changes, such a difference in sign seems, at first sight, unexpected. Moreover, in correlating these phenomena with those of the Volta contact effect between metals in air, it is probable that there will at all events be a general agreement between them, and it seems unlikely that metals would show a difference of sign in their potential differences with air, if that difference is due to actual oxidation, or to an affinity which tends to oxidation. On the other hand, Nernst's theory of electrolytic solution pressure offers a possible explanation of the difference in sign as given above. Whatever be the final outcome of the problem, we may take Neumann's numbers and similar results as true relative values, though eventually a constant error may have to be added to or subtracted from them.

The deposition and solution of metals from solutions of their salts have been shown to be reversible processes, the slightest change of electromotive force being enough to reverse the operation. Hence Neumann's results give both the electromotive force set up by the solution of such a metal as zinc, and the decomposition voltage needed to deposit it from a solution of the same concentration.

The potential differences rise with the dilution of the solution, and it seems not unlikely that some approximation to the logarithmic formulae of concentration cells may hold also in the present case. If so, the potential difference would assume a very high value for a very dilute solution.

Electrolysis has long been used to separate metals from each other. The theory of this process will now be clear. Let us suppose that we have a mixed solution of zinc and copper sulphates. The deposition point of copper is $-0.515$ volt, and that of zinc $+0.524$ volt. Thus, if the total electromotive force applied be enough to give a potential difference at the cathode greater than $-0.515$ volt but less than $+0.524$ volt, copper only will be deposited, for although its deposition point rises as the amount of copper gets less, this change is very small, and all traces of copper which could be detected by chemical analysis will be removed from the solution before the deposition point rises to that of zinc. If the electromotive force at the cathode be now increased above $+0.524$ volt, the zinc likewise can be removed from solution.

Even without this adjustment of electromotive force, if the solution be kept well stirred to prevent the local exhaustion of one metal at the electrodes, complete separation can nearly be effected. For, as we have seen, as long as there is any of the metal of lower deposition point present, none of the other is liberated. This principle is used in a process of copper refining. A plate of pure copper forms the cathode in a bath of copper sulphate. The anode is a thick plate of impure copper, probably containing metals both less and more easily deposited than copper. The bath is stirred, and when the current flows, copper and all more oxidizable metals are dissolved, while the less oxidizable metals, such as gold and silver, fall to the bottom of the vessel; for while copper is present in excess the current will dissolve it rather than more resisting metals. In the neighbourhood of the cathode, however, there will be a large excess of copper together with other metals, such as zinc, more easily oxidizable and therefore of higher deposition points. As long as any copper is near, therefore, none of the other metals are deposited, and pure copper is obtained at the cathode.

On the other hand, by increasing the current density, it is usually possible to exhaust the one metal from the layers of solution next the electrode faster than either stirring or diffusion will replace it. The other metal must then also be used by the current, and, by proper adjustment of conditions, it is possible to deposit alloys, the percentage composition of which can be altered by varying the current density.

For many purposes in chemical and biological research it is necessary to measure the concentration of hydrogen ions in a solution. This is done by means of a hydrogen electrode, consisting of a metal such as platinum or a platinum-iridium alloy saturated with occluded hydrogen, and used in combination with a calomel electrode to form a cell. The electromotive force of this cell gives the hydrogen ion concentration in the liquid in contact with the hydrogen electrode. Such a cell is illustrated in Fig. 109a. The hydrogen electrode is provided with a small hollow platinum cylinder, perforated by a number of minute holes through which hydrogen is blown into the solution to be examined, thus meeting the platinum saturated and the solution stirred. The calomel electrode is connected with the solution to be examined by a tube filled with a solution of potassium chloride run in through a stop-cock from a reservoir. The E.M.F. is measured by means of a potentiometer, and will be found to rise as the alkalinity of the solution round the hydrogen electrode increases, that is, as the hydrogen ion concentration diminishes.

If a deci-normal solution of hydrochloric acid were completely ionized, there would be 0·1 grm. equivalent of hydrogen ions per

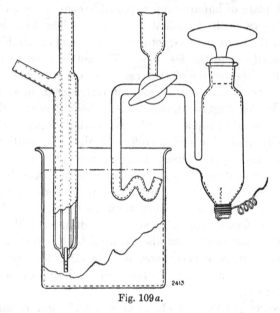

Fig. 109 a.

litre. As at 18° C. only 0·84 is ionized, the hydrogen ion concentration is $8·4 \times 10^{-2}$. A more convenient way of expressing this concentration is to take the logarithm to the base 10 of the concentrate with its sign reversed as $P_H$, the hydrogen ion exponent. Now $\log_{10} 8·4 \times 10^{-2} = 0·92 - 2 = -1·08$ and the value of $P_H$ is 1·08. Pure water at 21° C. has an ionic concentration of $10^{-7}$, for which $P_H$ is 7. If $P_H$ is greater than 7, the liquid is alkaline.

The hydrogen ion concentration may also be estimated by its effect on the colour of one of a series of indicators which may be added to the solution. Either of these methods gives the actual value of the ionic concentration at the moment, whereas titration with an alkali gives the total amount of acid whether ionized or un-ionized. Thus titration of a normal solution (that is a solution containing one germ-equivalent per litre) of a "weak" acid, such as acetic, gives the same acidity as of a normal solution of a strong acid, such as hydrochloric. But measurement of the hydrogen ion concentration shows that it is far greater for hydrochloric than for acetic acid.

**94.** Another set of electrocapillary phenomena, like those
Electric considered above, probably depend on the natural
endosmose. potential differences at the surface of separation of
two unlike substances—in this case an electrolyte and an insulator.
If an electric current be passed through a vessel divided into two
compartments by means of a porous partition and filled with some
solution, we shall find that, in general, besides the changes in concentration at the electrodes which were described on p. 238 under
the head of migration, there is a bodily transfer of the liquid, usually
in the direction of the current, through the porous plate. To this
phenomenon the name of electric endosmose is given.

If the pressure be kept constant on both sides of the partition,
the volume of liquid which flows through, as measured by the overflow, is proportional to the total electric transfer, and is independent
of the area and thickness of the plate; it varies much with the
nature of the solution, being greater with liquids of high specific
resistance, and, in solutions of different concentrations of any one
substance, is approximately proportional to the specific resistance.

If the liquid be not allowed to overflow, the pressure on one
side of the porous wall will increase. The final pressure is directly
proportional to the electromotive force between the faces of the
partition, and therefore to the current through it; for a given
current it varies inversely as the area of face of the porous wall
and directly as its thickness. In this case, the flux of liquid due
to the electric forces must be equal and in the opposite direction to
that caused by the difference of hydrostatic pressure. Considering
the porous wall to consist of a collection of capillary tubes, we may
apply the laws of flow through such tubes to the reverse flux under
the hydrostatic forces, and this explanation has been supported
by Quincke, who proved that the pressure produced by electric
endosmose through a capillary glass tube was inversely proportional
to the fourth power of the diameter of the tube. The pressures were
considerable with distilled water, but ceased to be perceptible with
liquids of high conductivity such as solutions of salts and acids.

**95.** Throughout our investigation of the electrical properties
The theory of solutions we have constantly been led to infer that
of electrolytic
dissociation. the ions of electrolytes are to a certain extent inde-

pendent of each other. If we eliminate the polarization at the electrodes, the flow of the current is in accordance with Ohm's law, and therefore, any electromotive force, however small, will produce a corresponding current. Hence there can be no appreciable reverse electric forces to be overcome in the interior of the electrolyte. The applied electromotive force, then, simply moves the ions against the frictional resistance of the liquid, it does no work in separating them from each other. We see that the electric properties of solutions indicate a freedom of interchange between the parts of the dissolved molecules—a freedom also suggested by the phenomena of chemical double decomposition.

The existence of specific coefficients of mobility, as characteristic properties of certain ions in very dilute solutions, involves the idea of independent migration, and suggests that the freedom of the ions from each other persists during the greater part of the time, and is not merely a power of interchange at the moments of molecular collision. If it were only a momentary freedom, the convective passage of the ions in opposite directions through the liquid, indicated by Faraday's law, would be explained by a continual handing on of the ions from molecule to molecule. The ions would work their way along by taking advantage of the intermolecular collisions, and the ionic velocities would depend on the frequency of these collisions; a frequency which, as indicated by the kinetic theory, depends on the square of the concentration. Now, as we saw on page 240, the conductivity of a solution varies as the product of the concentration and the relative ionic velocity; on this view, then, the conductivity will be proportional to the cube of the concentration. The facts described on page 236 do not bear out this result. In dilute solutions of water at all events, the conductivity is proportional to the concentration, and, as the concentration rises, the conductivity increases at a slower rate. It is difficult to see how these relations could hold except as a consequence of an almost complete migratory freedom of the ions of dilute aqueous solutions, and very strong evidence is thus obtained in favour of a theory of ionic dissociation.

Preconceived ideas would not, perhaps, lead us to expect that substances, which, like the mineral salts and acids, show great chemical stability when solid, should be dissociated into their ions almost completely when dissolved in water. It must, however, be remembered that it is precisely these bodies which possess the greatest chemical activity, that is to say, most readily exchange their parts with those of other substances. That a solution of hydrochloric acid, again, does not exhibit the properties of dissolved hydrogen and chlorine, though it has been urged as an objection, is not a valid argument against the theory of dissociation, for the ions are certainly in conditions differing from 'those in which the atoms of the same elements exist in their usual state. Whether or not there is combination between the ions and the solvent, and whatever be the exact relation between the ions and the charges they carry, we are certain that a definite quantity of electricity has to pass between an ion and the electrode before the substance can be liberated in a normal chemical state, say as gaseous hydrogen or chlorine. The energy associated with a substance when ionized must therefore be very different in quantity and character from that associated with it when in its normal chemical condition, and there is no reason to assume identity of properties in the two states.

It has been suggested that, if really dissociated from each other, the two ions of a dissolved salt would generally diffuse at different rates, and therefore ought to be separable. If such separation occurred to any large extent, however, electrostatic forces between the ions would arise and increase till further division was prevented. Nevertheless, some separation should undoubtedly occur, and, as a matter of fact, a volume of water in contact with the solution of an electrolyte is found to take, relatively to the solution, a potential of the same sign as the charge on the ion which has the greater mobility and therefore the quicker rate of diffusion.

The rate of diffusion may be calculated by treating the osmotic pressure as the driving pressure, and reckoning the speed attained from the known mobilities of the ions under electrical forces. The values of the diffusive constants and potential differences thus deduced agree well with those found by experiment.

The dissociation required by the theory is a separation of the ions from each other, securing complete migratory independence. There is nothing to suggest that the ions are free from all chemical combination. The hypothesis of electrolytic dissociation is entirely independent of any particular view as to the nature of solution, which may well be chemical in its fundamental processes. All that is required to interpret the electrical phenomena is the freedom of the migrating ions from each other; they may quite possibly be combined in some way with the solvent.

In some cases, particularly in solvents other than water, it seems necessary to assume that the solute exists in solution as double or treble molecules, which, by dissociation, give rise to electrified ions. We may, on the other hand, have double molecules formed between one molecule of the salt and one or more of the solvent, structures such as $NaCl \cdot 2H_2O$, which then dissociate into $\overset{+}{Na} \cdot H_2O$ and $\overset{-}{Cl} \cdot H_2O$. Evidence is now accumulating, however, which seems to show that the ions of electrolytic solutions are to be imagined as composed of a central charged nucleus (the Na or Cl in the case taken) round which are collected numbers of solvent molecules, forming an inclosing envelope which moves with the nucleus, and forms an essential part of the ion. Thus Kohlrausch found that the variation with temperature of the mobilities of different ions was similar to the variation with temperature of the fluidity (*i e.* the reciprocal of the viscosity) of water. This suggests that the friction on the moving ions is the friction of water on water in bulk. Bousfield and Lowry have treated the movement of the ions as similar to the motion of a small sphere through a viscous fluid, and have applied Stokes' formula to calculate the average radius of the sphere consisting of the central charged nucleus and the envelope of solvent. The calculated value of this radius depends on the temperature, the concentration, and the other conditions of the solution.

Such hypotheses as to the nature of the ions do not affect the main conception of the dissociation theory, which holds that the parts of the electrolyte (the Na and Cl in our example) are independently mobile in the solution. So much seems to be demanded by the electrical evidence, and by certain osmotic phenomena which we must now describe.

**96.**   Certain membranes, made by depositing insoluble chemical Osmotic precipitates within the walls of porous earthenware pressure. cells, are found to be permeable to solvents, and impermeable, or much less permeable, to certain substances when dissolved therein. In a closed cell with such a semipermeable wall, the presence of a solution will cause solvent to enter from without, and a pressure will be set up within, which finally reaches an equilibrium value known as the osmotic pressure of the final solution.

Whatever view we take of the fundamental nature of a solution, we must imagine the dissolved substance scattered as a number of discrete particles throughout the volume of the solvent. The nature of the interaction which occurs between the solute and the solvent is unknown, possibly unknowable; but, whatever it may be, each particle of solute will affect only a minute sphere of solvent lying round it. The solution, then, may be regarded as containing a number of little systems, each composed of a solute particle surrounded by an atmosphere of solvent in some way influenced by its nucleus.

While the solution is concentrated, the little spheres will intersect each other, and the addition of further solvent will involve some change in the interaction between solute and solvent. But, in the process of dilution, a time will come when the spheres are beyond each other's reach, and the addition of more solvent merely increases their mutual separation without affecting their internal structure.

Thus, in a dilute solution, the energy-change of further dilution is merely the energy-change involved in separating the particles of the solute further from each other; it will not depend on the nature of any possible interaction between the solute and the solvent. The change of energy is thus independent of the nature of the solvent, and will be the same whether that solvent be water, alcohol, or any other liquid. It will even be the same when, in cases where that is possible, the solvent is removed altogether, and the solute is obtained in the gaseous state.

If we imagine that the bottom of a frictionless engine cylinder is made of a semipermeable membrane, separating a solution within the cylinder from a solvent without, it is easy to see that osmotic pressure may be made to do work, which will be measured by the pressure multiplied by the change of volume. Thus the osmotic pressure is measured by the change of the available energy per unit increase of volume; that is, by the rate of change with volume in the available energy of dilution.

In this manner we arrive at the conclusion that the osmotic pressure must be equal in amount to the gaseous pressure exerted by the same number of molecules when vaporised, and must conform to the laws which describe the temperature, pressure, and volume relations of gaseous matter. The result is seen clearly to be independent of any hypothesis concerning the mechanism of the pressure or the nature of the solution.

By the principles of thermodynamics, a connexion may be traced between the osmotic pressure of a solution and the difference between the freezing and boiling points of the solution and of its pure solvent. Thus, the number of pressure-producing particles in a given solution may be determined by measuring its osmotic pressure, the depression of its freezing point, or the rise of its boiling point compared with those of the solvent.

In certain dilute solutions, such as those of sugar in water, the gaseous value for the osmotic pressure has been verified by direct measurement. This is confirmed also by observations on the freezing points, and, in this case, it is easy to extend the experiments to solutions of electrolytes which appear to leak through semipermeable membranes. The freezing points of very dilute aqueous solutions of electrolytes indicate a number of pressure-producing particles greater than that suggested by the ordinary molecular weight of the salt, and, for the most dilute solutions examined, greater in the approximate ratio of the number of ions into which the salt must be imagined to be dissociated in order to explain the electrical relations.

Thus, in dilute aqueous solutions the evidence goes to show that certain solutes are dissociated, and that the dissociation is the same in kind as that required to explain the electrical phenomena. In other solvents dissociation into non-electrified parts seems sometimes to occur, for cases are known where the osmotic effects indicate dissociation, but no electrical conductivity exists. For solutions in water, however, the same kind of dissociation is indicated by the two lines of research. It is, perhaps, advisable to repeat that the dissociation needed is the separation of the ions from each other, not necessarily from the solvent; the dissociated ions may well be linked with solvent molecules, or act as nuclei at the centres of watery envelopes as already suggested.

**97.** A relation has been traced between the coefficients of
Chemical         chemical affinity of solutions of acids and bases and
relations.         their electrical conductivity. It seems likely that, in
dilute aqueous solution, the electrical ions of the solute alone are
concerned in quick chemical interchanges, though here, again, rapid
changes seem to occur in other solvents with solutions which show
no conductivity.

If dilute solutions of acids and alkalies are to be regarded as
containing only dissociated ions, the process of neutralization of
hydrochloric acid and potash will be represented by the equation

$$\overset{+}{H} + \overset{-}{Cl} + \overset{+}{K} + \overset{-}{OH} = \overset{+}{K} + \overset{-}{Cl} + H_2O.$$

The resultant water, in presence of an excess of its own substance,
is dissociated only to a very slight extent, and we see at once the
explanation of the observed fact that equivalent quantities of all
strong acids and alkalies evolve the same amount of heat when
reacting with each other: the only chemical change consists in the
formation of a molecule of water from its ions, and this holds whatever be the acid and álkali, provided they are highly dissociated.

If we assume that the solution of an electrolyte is so dilute that
the solute particles and their attendant watery atmospheres are
beyond each other's influence, the number of molecules dissociating
into their ions per second will be proportional to the concentration
$c_1$ of the undissociated molecules. Two ions will recombine in a
certain fractional number of the cases of collision between two
opposite ions. The number of recombinations per second will
therefore be proportional to the product of $c_2$ the concentration of
one ion and $c_3$ that of the other. These concentrations are equal,
and the rate of recombination varies as $c_2^2$.

When equilibrium is reached, the rates of dissociation and recombination will be equal, and we have the so-called mass law of
chemical action

$$k_1 c_1 = k_2 c_2^2.$$

If $\alpha$ be the coefficient of ionization, and $v$ the volume of the
solution containing one gramme-molecule of solute, $c_1$ will be
$(1-\alpha)v$, and $c_2$ will be $\alpha/v$. Hence

$$\frac{\alpha^2}{v(1-\alpha)} = \frac{k_2}{k_1} = \text{constant.}$$

This result, known as Ostwald's dilution law, is true for weak electrolytes such as ammonia or acetic acid, but fails for strong electrolytes such as potassium chloride or hydrochloric acid. Probably the extended influence of charged ions, compared with that of neutral molecules, causes the attendant watery atmospheres of ions to be so large that the solutions cease to be effectively "dilute" even at very small concentrations.

In some cases, as for instance in the solutions of some organic substances in the liquefied hydrides of chlorine, bromine and iodine studied by B. D. Steele, not only does this dilution law fail, but the equivalent conductivity actually increases with the concentration. This phenomenon may be explained on the supposition that the simple molecule $AB$ of the solute cannot be dissociated into electrical ions, but that polymeric molecules such as $A_nB_n$ can be so dissociated. In accordance with the mass law of chemical action the greater the concentration the more polymeric molecules are formed, and, therefore, the greater the fraction of the solute subject to ionization.

**98.** Certain non-crystalline or colloidal substances, such as ferric hydroxide, arsenious trisulphide, and gold, silver or platinum, can be prepared in liquids in the form of finely divided suspensions which possess some of the properties of electrolytic ions. The particles in these colloidal solutions travel under an electromotive force with velocities about equal to those of slow-moving ions, the direction of motion depending on the nature of the substance and of the solvent. This result indicates that the colloid particles carry electric charges.

The electrical properties of colloidal solutions.

The charges may be neutralized by the addition of electrolytes, which also coagulate the colloid solution, the coagulation being most rapid when the neutralization is complete. These changes seem to be accompanied by the absorption by the particles of some of the ions of the electrolyte of opposite electric sign. For ions of different chemical valency, the amounts so absorbed are equivalent, but the concentrations of electrolyte in the solution necessary to produce coagulation, and the reciprocals of those concentrations the coagulative powers, depend, in a very remarkable way, on the chemical valency of the ion with opposite charge to that on the colloid particles. The averages of the coagulative powers of salts of univalent, divalent, and trivalent metals are found to be proportional to the numbers $1 : 35 : 1023$ respectively. An explanation of these phenomena has been offered by the present writer.

Let us suppose that, in order to produce the aggregation of colloidal particles which constitutes coagulation, a certain minimum electric charge has to be brought within reach of a colloidal group, and that such conjunctions must occur with a certain minimum frequency throughout the solution. Since the electric charge on an ion is proportional to the valency of the ion, we shall get equal charges by the conjunction of $2n$ triads, $3n$ diads, or $6n$ monads, where $n$ is any whole number.

The chance conjunctions of a large number of particles moving like the ions of an electrolytic solution can be investigated by the principles of the kinetic theory of gases. If $1/x$ denote the chance of one ion colliding with a colloidal particle, the chance that two ions should collide with it is the product of their separate chances, or $1/x^2$, and so on. When applied to the case in hand, these principles lead to the conclusion that the relative coagulative powers of univalent, divalent, and trivalent ions will be proportional to the ratios $1 : n : n^2$. The value of $n$, which depends on a number of unknown factors, remains arbitrary. It we assume that $n$ is 32, $n^2$ is 1024, and we get the numbers $1 : 32 : 1024$ to compare with the experimental values of the relative coagulative powers $1 : 35 : 1023$. The effect of other factors, such as the influence of the opposite ion, are omitted in this theory, but it seems to give a preliminary account of the subject with some measure of success.

The fundamental principles of the theory of electrolytic dissociation are able to explain and coordinate many different groups of phenomena. To what extent the theory is applicable to solutions in solvents other than water, or to fused or solid electrolytes, is a problem now under investigation in many physical and chemical laboratories.

### REFERENCES.

*The Theory of Solution*; by W. C. Dampier Whetham. Chapters VIII to XIV.
*Electro-chemistry*; by R. A. Lehfeldt.

# CHAPTER XI

## THE CONDUCTION OF ELECTRICITY THROUGH GASES

Introduction. Electric discharge through gases at low pressure. Röntgen rays. Cathode rays. Canal rays. Ionization of gases. Conductivity of gases. Mobilities of gaseous ions. Ionization by collision. Gaseous ions as condensation nuclei. Diffusion of ions. The nature of gaseous ions. Ions from incandescent bodies. The corpuscular theory of conduction.

**99.** UNLIKE the liquid solutions and other electrolytes studied in the last chapter, gases, in normal conditions, are almost perfect insulators of electricity. Usually the leakage of charge from a body suspended in air is due mainly to a slight conduction along the supports. For a long time it was a subject of controversy whether or not any part of the leakage took place through the air. Although this question is now answered in the affirmative, the conductivity of gases is, in general, extremely small; it may, however, be increased greatly by the passage of Röntgen rays, the incidence of ultra-violet light on a metal plate, the neighbourhood of flames, of incandescent metals, or of radio-active substances.

Introduction.

Apparently the first to notice the electric spark was Otto von Guericke (1602—1686) of Magdeburg, the inventor of the air-pump. From his time forward the spark continued to excite the curiosity of investigators, and, as we have seen, Franklin proved its identity with the lightning flash.

By means of the spark, electrified bodies are discharged almost instantaneously, showing that a path of high conductivity is opened thereby. This fact is demonstrated in another striking manner by the phenomena of the electric arc, first described by Davy in 1801. If the terminals of a battery of low resistance, giving 60 to 80 volts difference of potential, be connected with rods of hard gas carbon, an intensely luminous arc is started by touching the carbons and then drawing them a short distance apart. Automatic devices, which keep the carbons at a constant distance as their ends are consumed, are used in various kinds of arc lamps. The currents through arcs may rise to several amperes, and the conductivity is of the same order as that of metallic conductors. The phenomena of the spark and of the arc show that, in normal conditions, gases withstand a certain electric intensity, but that their resistance breaks down under a definite electric stress, and remains low if the current which then passes be maintained. This behaviour should be compared with that shown by metallic conductors conforming to Ohm's law; in their case the smallest electromotive force produces a corresponding current, which increases in direct proportion to the applied electromotive force. Again, in order to pass a current through an electrolytic solution between metal plates, we must apply an electromotive force greater than the reverse force of polarization at the electrodes. But, in that case, the electromotive force needed to maintain the current is equal to that required to start it; the passage of the current does not modify the physical state of the medium.

**100.** The possible length of spark, with a given difference of potential, increases as the density of the gas is diminished by means of an air-pump. The relation between pressure and sparking potential was first investigated systematically in 1834 by Harris, whose experiments were confirmed and extended by Faraday in 1838. As exhaustion proceeds, the sparking potential diminishes, reaches a minimum, and then increases again very rapidly as the pressure sinks to a small fraction of a millimetre of mercury. In a "good vacuum," it is difficult to get any discharge to pass. The minimum sparking potential occurs at a pressure which depends on the length of spark-gap used. The critical pressure increases as the length of gap diminishes. The product of these two quantities is very nearly constant; it is independent of the nature of the electrodes, and thus may be regarded as a characteristic property of the gas.

Since the critical pressure increases as the length of spark-gap diminishes, for very short spark-gaps the critical pressure is fairly high. Hence, at very low pressures, it is easier for a spark to pass by a long path through a gas than by a short one. This curious result shows that the passage of the spark is facilitated by the presence of more gas between the electrodes, in fact it has been found by Paschen and by Carr that the sparking potential depends only on the product of the pressure and the spark-length—that is, upon the mass of gas between unit area of the electrodes.

Fig. 110.

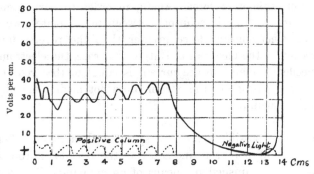

Hydrogen.  Pressure 2·25 mm.  Current 0·586 m.a.

Fig. 111.

Very beautiful appearances are presented by the discharge of an induction coil through gases at low pressure. The general type of discharge shows the following sequence of phenomena as the tube is exhausted. The first discharge consists of a luminous line down the middle of the tube. This line then broadens out till it seems to fill the tube. At pressures equivalent to that of about half a millimetre of mercury, a series of fluctuating striations, known as the positive column, appears in the neighbourhood of the anode (*a*, Fig. 110). A narrow dark space, called Crookes' dark space, surrounds the cathode, and this is separated from a second dark space, called after Faraday, by what is known as the negative glow.

By introducing platinum wires at different positions along the discharge, the distribution of potential has been investigated. Figure 111 shows a curve giving the relation between the distance from the anode and the potential-gradient in volts per centimetre. The difference of potential between the cathode and the negative glow is independent of the pressure of the gas or the distance between the electrodes. This potential-difference has been shown by R. J. Strutt to be equal to the minimum sparking potential.

If the length of the tube be increased, it is the positive column alone which increases with it; the two dark spaces, and the negative glow, vary very little with the length of the tube.

The effect of very high vacua on the electric discharge was first investigated systematically by Sir William Crookes. As the air is removed, it is found that the dark space nearest the cathode gradually extends, until eventually it fills the whole tube. At this stage, green phosphorescent effects begin to appear on the anode and on the glass opposite the cathode. If a solid object, such as a screen of mica, be interposed between the glass and the cathode, a sharp shadow is seen, showing from its position that rays capable of producing phosphorescence proceed in straight lines from the cathode. These cathode rays possess energy, for a light windmill placed in their path will rotate; moreover, they are deflected by a magnet, in the same direction as would be negatively electrified particles, travelling in the course of the rays. For this reason, Crookes and other English observers from the first contended that the cathode rays were to be regarded as a flight of negatively electrified material particles; while, on the contrary, it was believed for some time in Germany, where many experiments were also made, that the cathode rays, like those of ordinary light, were of the nature of aethereal waves.

**101.** In the year 1895, Professor Röntgen of Munich noticed that photographic plates became fogged, as though they had been exposed to light, if they were kept under cover in the neighbourhood of a highly exhausted tube through which electric discharges from an induction coil were passing. He investigated this effect, and found that, when cathode rays impinged either on the glass of the tube, or on the anode, or on any metallic

Röntgen rays.

plate within the tube, a type of radiation was produced which would penetrate many substances opaque to ordinary light. Dense bodies, like metal or bone, absorbed the rays more fully than did lighter materials, such as leather or flesh; and Röntgen, at once putting this discovery to some purpose, was able to photograph the coins in his purse and the bones in his hand. Moreover, these Röntgen or X-rays produce phosphorescence on screens of barium platino-cyanide and other similar salts, and, by using these screens in place of a photographic plate, objects, usually hidden from our eyes, may be made visible.

The form of apparatus usually employed for the production of Röntgen rays is shown in Fig. 112. The cathode is made of a concave aluminium cup which focuses the cathode rays on to a second aluminium plate or anti-cathode, fixed near the middle of

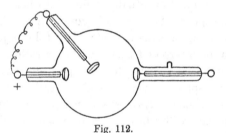

Fig. 112.

the exhausted bulb. Thus the anti-cathode is bombarded with cathode rays and becomes the source of X-rays. The anode may be placed anywhere in the bulb. It is usual to connect it with the anti-cathode. The character of the rays depends on the degree of exhaustion of the bulb in which they are produced. The higher the exhaustion, the more penetrating the rays become. The very "hard" rays obtained with high exhaustions are absorbed by different kinds of matter in almost exact proportion to the density of the matter. This relation does not hold for the "softer" rays given by bulbs containing air at higher pressures.

For several years after their discovery, the physical nature of the X-rays was widely discussed, and for a long time no general concensus of opinion was reached. Their photographic effects and the fluorescence they produced on suitable screens suggested that, like ordinary light, they were to be regarded as waves in the luminiferous aether. The power they possess of penetrating some opaque substances does not forbid such an assumption, for a difference in the wave-length, that is, in the period of vibration, is sufficient to produce marked differences in the penetration of ordinary light. Glass, transparent to the visible rays, is opaque to those invisible rays of longer wave-length which possess great heating power—hence its use in fire-screens; while a solution of iodine in bisulphide of carbon is opaque to luminous radiation, but allows the long waves to pass.

Röntgen rays are not refracted like ordinary light, and very little trace of regular reflection has been detected. Moreover, it is only with great difficulty that they have been made to show signs of such a typical property as polarization. Two plates of tourmaline seem to be as transparent to the rays when the axes of the crystals are crossed as when the axes are parallel. Such indications show that if there be identity in nature between Röntgen rays and ordinary light, the wave-lengths must be very different.

On the other hand, the rays suffer no deviation when acted on by a magnetic or by an electric field of force, a result which indicates that they are not projected particles carrying electric charges.

On Maxwell's theory, now universally accepted, light is explained as a series of electro-magnetic waves. But, as we have said, Röntgen rays are produced when a cathode ray strikes a solid object; and, if we take the cathode rays to be streams of electrified particles, it may be shown that electro-magnetic disturbances will be started by their impact.

In the year 1896 Sir George Stokes suggested that an explanation should be sought in the hypothesis that X-rays were single pulses travelling through the aether with the same velocity as light, but not followed by a train of waves, the thickness of the pulse, in which the whole disturbance is concentrated, being considerably smaller than the wave-length of any visible light.

But in more recent years X-rays have been shown to be very short aethereal waves, the wave-length being of the order of $10^{-8}$ centimetre.

In 1913 Lane suggested that the atomic structure of crystals might diffract X-rays, just as a finely ruled grating diffracts ordinary light, and give something in the nature of a spectrum. This theory has been verified by Friedrich and Kipping, and by Sir William Bragg and his son W. L. Bragg. The latter observers have shown that the periodic structure in the crystals is due to successive planes which contain atoms. They have measured the wave-length of the rays (about $10^{-8}$ cm.) and the dimensions of the space-lattice in various crystals.

Examined in this way, X-rays are found to consist of a mixture of heterogeneous radiation of all wave-lengths within certain limits and of more intense radiation of definite frequency, characteristic of the material of the anti-cathode, superimposed upon it. It is the characteristic line radiation that gives the diffraction pheno-mena, and with it a most important discovery was made in 1913 and 1914 by H. G. J. Moseley. When the target bombarded by cathode rays was changed from one metal to another, Moseley showed that the frequency of the characteristic X-rays underwent a change of a simple kind. The square root of the frequency $\nu$ of the strongest line in the X-ray spectrum increases by the same amount as we pass from element to element in the Periodic Table. If $\nu^{\frac{1}{2}}$ be multiplied by a constant so as to bring the regular increase to unity, we get a series of atomic numbers which are quite regular from aluminium (13) to gold (79). Known elements correspond with all the numbers except three which represent three possible elements still undiscovered.

These remarkable relations are best explained on a theory which regards the atom as a positively electrified nucleus surrounded by negative electric particles or electrons. The atomic number is identified with the number of positive units of electricity contained in the atomic nucleus. In hydrogen this is unity, and it rises by one as we pass up the Periodic Table to successive elements.

**102.** The successful explanation of the production and pro-perties of Röntgen or X-rays is strong evidence in favour of the view that the cathode rays, from which they arise, are flights of electrified particles. Other evidence pointing to the same conclusion is also available, and this conception of the nature of cathode rays is now accepted universally.

Cathode rays or negative rays.

A direct demonstration of the negative charge associated with cathode rays was given by Perrin. He showed that, when the rays within the discharge bulb were deflected by a magnet so that they fell, through a small hole in a metallic screen connected with earth, on to an insulated metallic cylinder connected with an electrometer, a strong negative electrification was imparted to the system. When the rays fell on other parts of the bulb, this electrification was not observed.

A less direct but more interesting method of inquiry has been used by Sir J. J. Thomson. The cathode rays are subjected successively to an electric and to a magnetic field of force. From the observed deflections, it is possible to calculate the velocity of the hypothetical electrified particles, to determine the sign of the electrification, and to measure $e/m$—the ratio of the charge $e$ carried by each particle to the mass $m$ of the carrying particle.

Let us imagine such an electrified particle moving in a straight path with uniform velocity. Let us apply a uniform electric field of intensity $f$ at right angles to the direction of motion. The effect is clearly the same as the effect of gravity on a body projected horizontally. The mechanical force on the particle is $fe$, and the acceleration $\alpha$ at right angles to the original motion is given by $fe/m$; the particle describes a parabola. The displacement in the direction of the electric force is

$$s_e = \tfrac{1}{2}\alpha t^2$$

$$= \tfrac{1}{2}\frac{fe}{m}\, t^2.$$

During the time $t$, the particle moves with its original velocity $v$ in the direction normal to the electric force, and traverses, in that direction, a path $l$, which is equal to $vt$. Hence $t^2$ is $l^2/v^2$, and the displacement is

$$s_e = \tfrac{1}{2}\frac{fe}{m}\frac{l^2}{v^2}.$$

Thus the equation involves the velocity of the particle and the ratio $e/m$ of the charge to the mass.

Now, instead of an electric field, let us apply to the moving particle a magnetic force $H$ at right angles to the direction of motion. The particle with its charge $e$ moving with a velocity $v$ is equivalent to a current $c$ passing along a path of length $\delta l$, and we have

$$ev = c\delta l.$$

The mechanical force on a current when placed in a magnetic field of force $H$ is $Hc\delta l \sin \theta$ (p. 109); and if the magnetic force be at right angles to the direction of motion, so that $\sin \theta$ is unity, the mechanical force on the particle is $Hev$. This force, according to Fig. 56 on page 108, is at right angles both to the magnetic force and to the direction of motion. Unlike the mechanical force due to the electric field, however, the force due to the magnetic field does not act constantly at right angles to the *original* direction of motion. It acts at right angles to the direction of motion *at each instant*. This characteristic is a property of circular motion, like the motion of a planet round the sun, or that of a stone whirled by a string. The centrifugal force is, in such cases, measured by $mv^2/r$, where $r$ is the radius of the orbit. Hence

$$Hev = \frac{mv^2}{r},$$

or

$$r = \frac{m}{e}\frac{v}{H}.$$

In practice, only a small part of the circle is described, so that the path of the particle is fairly straight. Hence, by the well-known property of the circle, the deflection is

$$s_m = \frac{l^2}{2r}$$

$$= \frac{l^2 H}{2v}\frac{e}{m}.$$

In this manner we obtain a new relation between the two quantities $v$ and $e/m$. Thus, by observing the magnetic and the electric deflections in a given length of path of the cathode rays, both $v$ and $e/m$ may be determined in absolute units.

In applying these results experimentally, Prof. Thomson used the apparatus shown in Fig. 113. Two electrodes are fixed in a highly exhausted glass tube by means of platinum wires fixed through the glass. $C$ is the cathode, and $A$ the anode. Through the anode, and through a disc $B$ of similar form connected with the earth, are bored holes, about a millimetre in diameter. Thus beyond

the disc $B$ we have a narrow pencil of cathode rays, removed from
the action of the potential gradient of the discharge, and therefore
moving with uniform velocity. The rays fall on the screen at the
other end of the tube, and give rise to a bright phosphorescent
spot at $p$. In the middle of the tube are fixed the two parallel
plates $D$ and $E$, which may be connected with the opposite
terminals of a battery of voltaic cells in order to establish an
electric field between them. When the rays pass out from between
the plates, they leave the electric field and proceed in straight lines.

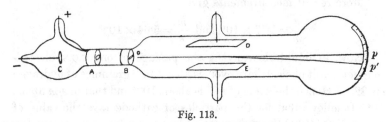

Fig. 113.

From the deflection $pp'$ at the end of the tube, the deflection $s_e$
in the field of force might be obtained. The method adopted,
however, was to apply simultaneously the electric field and a
uniform magnetic field, due to an external electro-magnet, which
produced a mechanical force on the rays in the opposite direction
to that due to the electric field. The magnetic field was adjusted
till no deflection occurred. Then the mechanical forces must be
equal and

$$fe = Hev$$

or

$$v = \frac{f}{H}.$$

From the displacement $s_m$ in a uniform magnetic field, the value
of $e/m$ was then obtained by means of a relation equivalent to the
equation given above,

$$s_m = \frac{l^2 H}{2v} \frac{e}{m}$$

or

$$\frac{e}{m} = \frac{2v s_m}{l^2 H}.$$

The results of these experiments show that the velocity of
cathode rays is not a constant. It varies even in the same discharge. This is shown by the fact that the magnetic field causes

the pencil of rays to spread out, so that the patch of light is larger
when deflected than when undeflected. The velocity varies still
more when the pressure or the nature of the residual gas is changed,
and values ranging from about $2 \times 10^9$ to $4 \times 10^9$ centimetres per
second have been obtained. On the other hand, the quantity $e/m$
appears to be constant. Whatever be the pressure or nature of the
gas, and whatever metal be used as electrode, the value of $e/m$ was
about $7·7 \times 10^6$, and that of $m/e$ about $1·3 \times 10^{-7}$ electromagnetic
units.

More recent measurements give

$$\frac{e}{m} = 1·77 \times 10^7 \text{ and } \frac{m}{e} = 5·64 \times 10^{-8}.$$

Now, in liquid electrolytes, the passage of one C.G.S. electromagnetic unit of electricity evolves about $10^{-4}$ grammes of hydrogen
(p. 229). Hence the value of $m/e$ is about $10^{-4}$ and that of $e/m$ about
$10^4$. It follows that for the particles of cathode rays the value of
$e/m$ is about 1800 times as great as, and the value of $m/e$ about the
1/1800th part of, the corresponding quantities for the hydrogen ion
in the electrolysis of liquids. We shall see later how to determine
the charge on gaseous particles similar to those of the cathode
rays. It will be found that the charge is identical with that on
a monovalent ion in a liquid electrolyte. The result stated above
then shows that the mass of a cathode ray particle is about the
1800th part of the mass of a simple hydrogen ion, that is of a
hydrogen atom. Thomson calls these particles by the Newtonian
name of corpuscles, and regards each corpuscle as a unit of negative
electricity. An atom with an extra corpuscle attached is a negatively
charged atom. An atom with one corpuscle less than its normal
complement is an atom with unit positive charge.

**103.** In an electric discharge through a rarefied gas, positive
Canal rays or     ions as well as negative help to carry the current.
positive rays.    The existence of these positive rays may be demonstrated by drilling a hole through the cathode and examining the
space behind it, that is, on the side of the cathode remote from
the anode. Phosphorescent effects which appear on the glass of
the vessel, or on a sensitive screen, make it clear that linear rays
pass through the hole. These "canal rays" were first discovered

by Goldstein in 1886, and their magnetic and electric deflections were examined by Weiss in 1898. He found that they contained positively electrified particles of the mass of chemical atoms, the minimum value of $e/m$ being about $10^4$, which is the ratio of the charge to the mass of the hydrogen atom in the electrolysis of liquids.

In 1910 and 1911, Sir J. J. Thomson carried much further the analysis of these rays. By using a large discharge tube, he was able to work at very low pressures, and, by fixing a long thin tube through the hole in the cathode, he obtained a very narrow pencil of rays. By the direct impact of these rays on a sensitive photographic plate, placed inside the apparatus, he found a most delicate method of recording them.

The rays were passed through coincident magnetic and electric fields, so arranged that their deflections were at right angles to each other. When undeflected, the rays strike a point $O$ on the screen on the axis of the cathode tube. When both magnetic and electric fields act, the rays impinge on a point $P$, where the length of the vertical line $PN$ is equal to the magnetic deflection, and the length of the horizontal line $ON$ to the electric deflection.

Now, as we saw on pp. 285, 286,

$$PN = A \frac{e}{mv} \text{ and } ON = B \frac{e}{mv^2},$$

where $A$ and $B$ are constants depending on the strengths of the magnetic and electric fields respectively, and on the geometrical data of the tube. Hence

$$\frac{m}{e} = \frac{A^2}{B} \frac{ON}{PN^2}.$$

If we have a type of carrier for which $m/e$ is constant, it follows that $PN^2/ON$ is constant whatever the velocity. Therefore, if we have a number of particles identical in mass and charge but differing in velocity, they will strike the plate in a series of points the locus of which is a parabola.

In accordance with this theory, when the photographic plate which has been exposed to the rays is developed, a series of parabolic curves is seen. If we know the constants $A$ and $B$, we can calculate the quantity $m/e$ for each curve from measurements of its position on the plate. We can compare the values of $m/e$ for two curves without determining $A$ and $B$; it is enough to measure $ON$ and $PN$. Hence such comparisons can be made from the photographs alone.

The lines which appear depend on the nature of the residual gas in the discharge tube. In hydrogen the fundamental line gives a value for $m/e$ of $10^{-4}$, the value of the hydrogen atom in liquid electrolytes. This line then is produced by a flight of particles which are hydrogen atoms carrying each a single ionic charge. Another line has a value for $m/e$ double that given above. It corresponds to the hydrogen molecule carrying a single charge.

The quantity $m/e$ is the characteristic constant for the electric carrier. When referred to its value for the single charged hydrogen atom as unity it is called by Thomson the "electric atomic weight." The following table gives the electric atomic weight and the probable corresponding carrier for the lines seen when nitrogen prepared from air is the residual gas.

Fig. 113 a.

| Positive | | Negative | |
|---|---|---|---|
| Electric Atomic weight | Carrier | Electric Atomic weight | Carrier |
| 1 | H + | 1 | H − |
| 1·99 | H$_2$+ | 11·2 | C − |
| 6·8 | N + + | 15·2 | O − |
| 11·4 | C + | | |
| 13·95 | N + | | |
| 28·1 | N$_2$+ | | |
| 39 | Ar+ | | |
| 100 | Hg + + | | |
| 198 | Hg + | | |

The symbol H + denotes an atom of hydrogen with one univalent ionic charge; N + + an atom of nitrogen with two such charges; and so on. Faint lines, with electric deflections opposite to those of the stronger lines, indicate that some carriers pass through the electro-magnetic field associated with negative charges, probably acquired by collisions with corpuscles behind the cathode.

The analogy between these photographs, and those obtained with optical spectra has led to the adoption of the name electric spectra in this case. Electric spectra possess certain advantages as instruments of research over their older prototypes. Firstly, the method is often more sensitive. Helium has been detected electrically, when no trace of its optical spectrum could be seen. Secondly, even transient or nascent substances are detected. Thirdly, the electric spectrum gives at once the electric atomic weight of the particles present, whereas long chemical and physical investigations are needed to determine the atomic weight of the particles giving an unknown line in an optical spectrum.

Thomson's results show that a gas through which an electric discharge passes is a very complex structure. Thus in oxygen at least eight different particles must be present besides negative corpuscles:

1. Ordinary molecular oxygen $O_2$.
2. Neutral atoms of oxygen O.
3. Atoms of oxygen with one positive charge O +.
4. Atoms of oxygen with two positive charges O + +.
5 Atoms of oxygen with one negative charge O −.
6. Molecules of oxygen with one positive charge $O_2$ +.
7. Molecules of ozone with one positive charge $O_3$ +.
8. Aggregates of six oxygen atoms with one positive charge $O_6$ +.

Another result of the experiments is to throw a very vivid light on the problem of the nature of the changes going on in the path of a discharge. Cathode rays produce positive ions by collision with particles of the gas near the anode. The positive ions begin to move under the action of the electric field, and become positive rays. Some of them pass through the tube in the cathode, and reach the photographic plate without change as primary canal rays. Others,

by collisions with corpuscles or atoms, lose their charges, and, flying
on with the velocity they then possess, suffer no deviation in their
path.  Of these last, some few atoms are capable of acquiring negative
charges by absorbing corpuscles in their way, incidentally showing
how great is the attraction between such neutral atoms and negative
electricity, since an atom moving with such great velocity is able
to pick up a negative corpuscle as it flies.

In all cases, it will be noticed, positive ions are structures
of atomic or molecular mass.  The sub-atomic particles, which
constitute the stream of cathode rays, are found associated with
negative charges only.  No corresponding positive corpuscle has
been discovered.  Positive electricity seems, as already indicated,
to consist in a defect in the number of negative corpuscles normally
associated with an atom or molecule.

In examining the element neon (atomic weight 20·2) Thomson
found two lines indicating weights of 20 and 22 respectively.  This
suggested that neon as ordinarily prepared might consist of a mixture

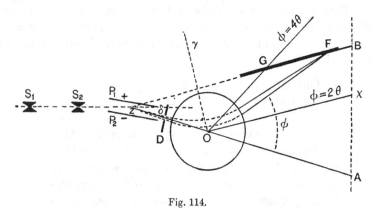

Fig. 114.

of two elements.  It will be seen later that the modern theory of
the atom, outlined on pp. 339, 340 gives an explanation of these
phenomena.  Such elements are known as isotopes and have now
been isolated in many cases by F. W. Aston.  Aston uses a modification of the apparatus.  Positive rays are sorted into a thin
ribbon by means of two parallel slits $S_1$, $S_2$ (Fig. 114), and then
spread into an electric spectrum by passing between two oppositely
electrified plates $P_1$, $P_2$.  A part of this spectrum is isolated by a

diaphragm $D$, and passed through a strong magnetic field which bends the rays back again through an angle larger than the electric deflection. The result is that rays having a constant value of $m/e$ will converge to a focus $F$, and by photography a spectrum is obtained dependent on mass and called a mass-spectrum.

Mass-spectra for neon and chlorine are given in Fig. 114$a$. The displacement to the right with increasing mass is roughly linear. The absolute values of the atomic mass are found by reference to the known lines of hydrogen, oxygen, etc. which usually occur as impurities. With this apparatus the isotopic nature of neon was confirmed. Then chlorine (35·46) was shown to consist of a mixture of 35 and 37. Lines due to hydrochloric acid, HCl, appear at 36 and 38, and many other elements have similarly been resolved into isotopes. Except hydrogen, 1·008, all elements examined have atomic weights that are exactly whole numbers on the scale oxygen = 16. This remarkable discovery of isotopes is confirmed by an independent line of research by Soddy, who has prepared lead by radio-active means differing in atomic weight from lead as usually obtained.

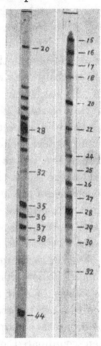

Fig 114$a$.

**104.** As already stated, in normal conditions, air and other gases are very good insulators; and an electrified body suspended in air will lose its charge only with extreme slowness if precautions be taken to prevent conduction along the supports. The gold-leaf electroscope shown in Figure 115, which is here reinserted from p 19 for convenience, contains an ingenious device, due to Mr C. T. R. Wilson, for preventing such leakage. The brass plate $D$, to which the gold-leaf $C$ is attached, is held in position by the bead of sulphur $A$, which is a very good insulator. The sulphur-bead is fused to a brass rod, which passes through the ebonite plug at the top of the instrument

*Ionization of gases at atmospheric pressure.*

and serves to suspend the sulphur-bead and the gold-leaf system.
The gold-leaf is charged by means of the moveable wire *EB*, which
may be brought into contact with the
plate *D* and then again removed. Any
leakage from the gold-leaf along the supports must take place over the sulphur-
bead *A* to the brass rod which suspends
the bead. Now, this rod may be connected
permanently with the pole of the battery
used to charge the wire *EB* and the gold-
leaf. Hence, when first charged, the gold-
leaf is at the same potential as its supports,
and there is no tendency for leakage to
occur. As the gold-leaf loses its charge
through the air, its potential sinks below
that of the supporting rod, and any leakage must tend to restore

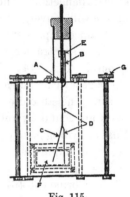

Fig. 115.

the charge of the leaf instead of to abstract it. While the gold-leaf
is near its original potential, the difference of potential between
it and its supports will be very small, and the usual slight leakage
is much reduced as well as reversed in direction.

Even with the precaution just described, the gold-leaf system
is found to lose its charge, though many hours may pass before
the charge is half dissipated. It follows that air in its normal
condition must possess some conductivity. The effect is so small,
however, that it is probable that its phenomena would never have
been coordinated or explained had we not possessed means of
increasing it at will. It has long been known that an electrified
body is discharged by contact with the hot gases of a flame, and,
soon after Röntgen's discovery of X-rays, it was noticed that an
electroscope lost its charge rapidly when the rays passed through
the air surrounding it. A similar result is obtained by the incidence
of ultra-violet light on a metallic surface, or by the neighbourhood
of radio-active substances. By any of these methods, the electrical
conductivity of gases may be increased very greatly.

The nature of the conductivity thus imparted is indicated by
the fact that a gas, which has been under the influence of one of
the agencies mentioned above, does not lose its newly acquired
properties immediately. It may be blown about from place to place,

and still will discharge an electroscope.  These effects may be investigated by an arrangement similar to that shown in Figure 115 $a$. The Röntgen rays are produced in a bulb placed within a box

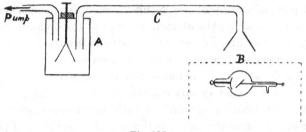

Fig. 115 $a$.

covered with thick lead except at the window $B$, through which alone the rays find an exit.  Thus the electroscope $A$ is screened from the direct action of the rays.  Through the walls of the electroscope two tubes pass, and, by attaching a water-pump to one of them, a slow current of air, drawn from the region traversed by the X-rays, may be sucked through the electroscope.

In these circumstances, the electroscope is found to lose its charge, while, if the current of air be stopped, or if the rays be cut off, the charge is retained.  If a plug of glass-wool be placed in the tube $C$, the electroscope is not discharged by the current of air drawn from the region subject to the action of the rays.  The conductivity of the air is also removed by bubbling the air through water, or by passing it through a metallic tube of fine bore.  The same effect is produced by sending the air through an electric field, which may be produced by means of a metallic tube of large bore, along the axis of which is stretched an insulated wire.  The tube and wire are connected with the opposite terminals of a battery of some hundreds of small accumulator cells, the tube being also joined to the earth.  After passing through such a tube, the conductivity of the air disappears.

These phenomena suggest that the conductivity of air is to be explained by the supposition of the presence of charged particles, somewhat of the same nature as the ions of liquid electrolytes.  In liquids, the ions are permanent, but in gases some ionizing agency is necessary for their production.  Moreover, when formed, the ions of gases are removed by contact with solids (glass-wool or the walls of a fine tube), by contact with liquids, or by the action of an electric field. Even without such processes, the ions gradually disappear spontaneously. Regarding the ions as oppositely electrified parts of molecules, it is natural to refer this disappearance to recombination under the attractive forces of the opposite charges. This hypothesis may be developed mathematically, and its consequences subjected to experimental examination.

Let $q$ denote the total number of ions, positive and negative, produced per second in one cubic centimetre of the gas by the action of the ionizing agency. If each uncharged molecule gives rise to one positive and one negative ion, the number $n$ of positive ions which exist at any moment in unit volume is the same as the number of negative ions.

Now, on the principles of the dynamical theory of gases, the number of collisions per second between opposite ions is proportional to the product of their concentrations, that is, to $n^2$. If a certain fraction $\beta$ of the collisions result in the formation of a neutral system, the number of ions which disappear per second in a cubic centimetre is $\beta n^2$. But $q$ is the constant rate of ionization, and the resultant total rate at which ions are formed is given by

$$\frac{dn}{dt} = q - \beta n^2.$$

In a steady state, when the ionization is constant, $dn/dt$ vanishes, and we have

$$q = \beta n^2.$$

If the ionizing agency be removed, $q$ becomes zero, and the rate of decay of the ionization is represented by the equation

$$\frac{dn}{dt} = -\beta n^2,$$

which, when integrated, gives

$$\frac{1}{n_0} - \frac{1}{n} = \beta t,$$

where $n_0$ is the initial number of ions, and $n$ the number after a time $t$.

The rate of recombination of gaseous ions may be investigated by an apparatus used by Sir E. Rutherford and shown in Fig. 116. A slow current of dry gas is drawn through a plug of cotton-wool to remove particles of dust, and then through a long metallic tube. At $T$, the gas is subjected to the action of an ionizing agency such as X-rays, or the radiations from a radio-active substance such as oxide of uranium. At different points along the tube, insulated electrodes, $A$, $B$, etc., are inserted, and, by connecting one of them

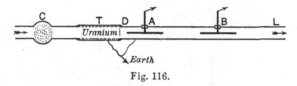

Fig. 116.

with one terminal of a battery of accumulators, and the earthed tube with the other terminal, the conductivity of the gas at the point may be measured. A gauze screen at $D$ shields the region near $T$ from the electric field, and serves to prevent the direct abstraction of ions by the electric force at $A$ or $B$. If the tube be of wide bore, the loss of ions caused by diffusion to the walls is small, and the decrease in conductivity of the gas is due to recombination alone. Knowing the velocity of the current of gas along the tube, the effect of time on the ionization may be investigated.

The results of such experiments agree well with the equations given above, and are thus consistent with the theory on which those equations are based. The quantity $\beta$ is found to be constant, and to have a value of about $3000e$, where $e$ is the charge on a gaseous ion. This quantity $e$ has been shown, by methods to be described hereafter, to be about $3\cdot4 \times 10^{-10}$ electrostatic units. Thus $\beta$ is about $10^{-6}$, and if $10^6$ ions be present per cubic centimetre, half of them recombine in about one second.

**105.** In metallic conductors the current is proportional to the
Conductivity of gases. electromotive force applied, in accordance with Ohm's law. In liquid electrolytes the same relation holds good, provided the effects of the polarization of the electrodes be eliminated. In order to examine the relation between the current and the difference of potential in gases, the apparatus indicated in

Fig. 117 may be employed.  The air between two parallel metallic plates $A$, $B$ is exposed to Röntgen rays, or to the radiation from some radio-active material.  One plate, $B$, is connected with one pair of quadrants of a quadrant electrometer (p. 46), the other pair of quadrants being connected with earth.  The second plate $A$ is joined to one pole of a battery of cells, of which the other pole

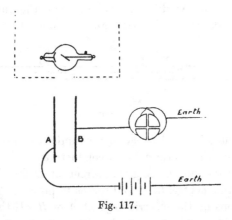

Fig. 117.

is put to earth.  At first, the opposite pairs of quadrants are connected together.  When they are disconnected, the conductivity of the gas enables the battery to charge up one pair of quadrants, and thus the rate of movement of the needle of the electrometer measures the current through the gas.  By changing the number of cells in the battery, the relation between the potential-difference and the current may be investigated.

For a small potential-difference, it will be found that a proportional current flows—the conduction conforms to Ohm's law, and, near the origin, the curve between $C$ and $E$ is a straight line.  As the potential-difference is increased, however, the current rises more slowly, till a limit is reached, and the curve becomes horizontal.  On the analogy with the shape of the curves of magnetization (p. 85) the current then passing through the gas is known as the saturation current.  It remains constant till the potential-difference is increased nearly to the value at which a spark passes, when a rapid increase of current again occurs.

Figures 118 and 119* show the relation between potential

* Rutherford's *Radio-activity*.

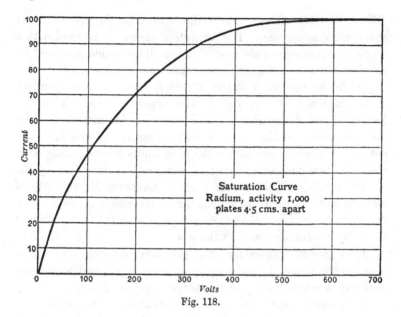

Fig. 118.

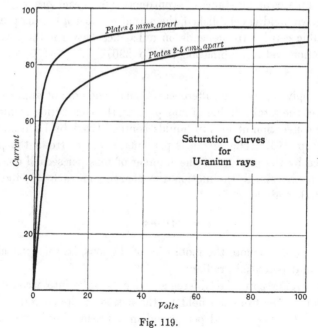

Fig. 119.

difference and current, when the ionization is produced by radio-active substances. The saturation current is reached with a much lower voltage in the case of the small ionization produced by uranium than with the large ionization due to radium.

While the current is at the saturation value, it is independent of the intensity of the applied electromotive force. This fact facilitates the determination of the relative ionization of a gas under different conditions. It is only necessary to know that the potential-difference lies within the wide limits corresponding with the saturation current, in order to measure the ionization by the current passing. The current may be measured by the rate of deflection of the needle of a quadrant electrometer, or by the rate at which the gold-leaf of an electroscope falls, as the charge thereof is lost by conduction through the gas.

If we assume that the ionizing agent acts uniformly everywhere between the plates, a condition approximately realised with some difficulty, a simple theory of this subject may be given.

Let $q$ ions per second be produced per cubic centimetre between parallel plates at a distance of $l$ centimetres from each other. When no electric field is established, the actual number $n$ of ions present per cubic centimetre depends on a balance between the rates of production and recombination, and (p. 296)

$$q = \beta n^2.$$

If we apply a potential-difference $E$, small enough not appreciably to affect the total number of ions present, the conduction is similar to the convection of ions in liquids contemplated by Kohlrausch's theory (p. 240). The current $c$ per square centimetre of the plate is given by the product of the number of ions per second crossing unit area normal to the motion, the charge $e$ on each ion, and the potential-gradient $E/l$. Hence

$$c = ne\,(u + v)\,\frac{E}{l},$$

where $u$ and $v$ denote the mobilities of the ions, i.e. their velocities under unit potential-gradient.

For higher potential-differences, we must take into account the removal of ions from the field by the action of the current. The number of ions produced per second in a prism of unit cross area and length $l$ is $ql$.  The maximum current $C$ will be obtained when all these ions are driven to the electrodes by the electric forces before appreciable recombination has taken place.  Thus

$$C = qle,$$

and the ratio of any very small current to the maximum current is given by

$$\frac{c}{C} = \frac{n\,(u+v)\,E}{ql^2}.$$

But $q$ is $\beta n^2$; hence, eliminating $n$,

$$\frac{c}{C} = \frac{(u+v)\,E}{l^2\sqrt{q\beta}}.$$

On the right-hand side of this equation, every term except $E$ is constant.  Hence, as experiment shows, a small current should be proportional to the applied potential-difference.

The saturation current,

$$C = qle,$$

increases with the rate of ionization, and with the distance between the plates.  Thus, with a constant high potential-difference, the current through a gas is increased by separating the plates, a paradoxical result, which is explained when we remember that, with a saturation-current, all the ions are used up as soon as formed; an increase of space between the plates then gives more room for the action of the ionizing agency, and increases the number of ions available.

**106.**  The sum of the opposite velocities of the complementary
Mobilities of   ions may be determined by a method based on the
gaseous ions.   equations just given.

If $u$ and $v$ denote the ionic mobilities of the positive and negative ions respectively, that is, the velocities with which they move under unit potential-gradient, the current density $c$, when it is small, is given by the equation

$$c = ne\,(u+v)\frac{E}{l},$$

where $n$ is the number of ions per unit volume, each ion carrying a charge $e$, and $E/l$ is the potential-gradient.  The current per unit area of two plates (Fig. 117) may be determined by observing the rate of deflection of the needle of an electrometer of known capacity. The quantity $ne$ may be found by a second experiment. The ionizing agent, let us suppose it to be Röntgen rays, is allowed to act till the gas has reached a steady state. Then, at the same instant, the rays are switched off, and a very high potential-difference is applied between the plates. All the ions which, at the instant, exist between the plates are swept to the electrodes before appreciable recombination can take place. The negative ions alone go to the plate $B$ connected with the electrometer, which thus receives a total charge equal to $nel$, where $l$ is the distance between the plates. This charge $Q$ may be measured in absolute units by the electrometer, and, combining the results of the two experiments, we have

$$\frac{c}{Q} = \frac{(u + v)\,E}{l^2},$$

whence $u + v$ may be calculated.

By this method Rutherford measured the sum of the ionic mobilities of different gases at atmospheric pressure. The gases were not specially dried.

| Gas | $u+v$ in cm./sec. | Gas | $u+v$ in cm./sec. |
|---|---|---|---|
| Hydrogen | ... 10·0 | Carbon dioxide ... | 2·15 |
| Oxygen ... | ... 2·8 | Sulphur dioxide | 0·99 |
| Nitrogen ... | ... 3·2 | Chlorine ... ... | 2·0 |
| Air ... | ... 3·2 | Hydrochloric acid | 2·55 |

The mobilities of the individual ions cannot be calculated by this method alone. The ratio of the velocities, which was determined for liquid electrolytes by Hittorf's migration experiments, must also be obtained.

A method of so doing, due to Zeleny, consists in finding the electric force required to push an ion against a stream of gas, moving with a known velocity in the opposite direction. The velocity of the negative ion is, in general, higher than that of the positive, the ratios being

| Gas | $v/u$ | Gas | $v/u$ |
|---|---|---|---|
| Air ... ... | 1·24 | Hydrogen ... | 1·14 |
| Nitrogen ... | 1·23 | Carbon dioxide ... | 1·00 |

By combining these results with those given by Rutherford, the individual mobilities may be calculated.

The individual mobilities have also been determined by direct observation in various ways. Zeleny again used a blast-method. He passed a stream of gas along a tube, and measured the distance that stream carried an ion forward while it was passing from the circumference of the tube to the axis under the influence of a radial electric force.

| Gas at 760 mm. of mercury pressure | Temp. in deg. Cent. | Velocities in cm./sec. per volt per cm. | | Ratio of velocities of negative and positive ions |
|---|---|---|---|---|
| | | Positive ion | Negative ion | |
| Air, dry ...   ...   ... | 13·5 | 1·36 | 1·87 | 1·375 |
| „  moist   ...   ... | 14 | 1·37 | 1·51 | 1·10 |
| Oxygen, dry   ...   ... | 17 | 1·36 | 1·80 | 1·32 |
| „   moist ...   ... | 16 | 1·29 | 1·52 | 1·18 |
| Carbon dioxide, dry   ... | 17·5 | ·76 | ·81 | 1·07 |
| „   „   moist ... | 17 | ·82 | ·75 | ·915 |
| Hydrogen, dry ...   ... | 20 | 6·70 | 7·95 | 1·19 |
| „   moist   ... | 20 | 5·30 | 5·60 | 1·05 |

It will be seen that, in most cases, these results agree with the sum of the opposite ionic mobilities given above. Other methods, also, give values in accordance with these figures.

It is found that, till the pressure sinks to a small fraction of an atmosphere, the velocity of an ion varies inversely as the pressure of the gas through which it moves. The resistance to the motion of the ion is thus proportional to the density of the gas, or to the number of collisions between the ion and the molecules of the gas. It is interesting to note that the ordinary viscosity of a gas is independent of the pressure. Thus the motion of a gaseous ion is not analogous to that of a large body moving through a viscous fluid, but to that of one molecule of a gas working its way among the others.

As the pressure sinks to that corresponding to about 10 millimetres of mercury or less, a rapid increase occurs in the mobility of the negative ion. No similar large change is observed for the positive ion. At these low pressures, some of the negative ions seem to resemble the minute structures detected in cathode rays. At higher pressures it is natural to picture the negative ion as composed of one of these corpuscles, with one or more neutral molecules attached.

**107.**  The cathode rays, as we have seen, consist of a flight of
Ionization   negatively charged particles.  When passing through
by collision.   a gas, these flying particles produce ionization, as
may be demonstrated by means of the apparatus of Figure 113 on
page 287.  If some residual air be left in the tube, and a difference
of potential be applied between the plates $C$ and $D$, no current
will pass between the plates till the cathode rays are turned on,
when a considerable current may be observed.  Similar particles
are emitted by some radio-active bodies, and these particles also
produce ionization when passing through a gas.

Such phenomena suggest that the flying particles ionize some of
the molecules of the gas by virtue of the collisions which must occur,
and this idea has been adopted and developed by many physicists.

When a potential-difference is established between two electrodes separated by a layer of gas, the few ions, which always seem
to be present, are put in motion.  The lower the pressure of the
gas, the longer is the free path of an ion before it collides with a
molecule of the gas, and thus the greater is the velocity acquired
by the ion, and the greater is the shock of a collision when it does
come.  If the pressure be so low that some of the negative ions are
free corpuscles, very high velocities will be reached between the
collisions.  In such cases, it is clear that the original negative ions
will produce new ions by collision, just as do the cathode rays.
These secondary ions may themselves produce tertiary ions, and so
on.  The result will be that a continually increasing supply of ions
is forthcoming, and eventually a continuous electric discharge passes
through the exhausted gas.

In this stage, the current is, of course, much greater than the
usual saturation current of Figures 118 and 119 on page 299.
These further phenomena have been studied from this point of
view by Townsend, who explains all spark-discharges by similar
considerations.  Figure 120 shows a complete curve between the
voltage as abscissa and the current as ordinate.  Here the saturation
current extends from 700 to 1100 volts, and a rapid increase is seen
beyond that point.

At atmospheric pressures, the potential-gradient required to
produce ions by collision and start the spark is very great—some
30,000 volts per centimeter.  The first spark facilitates the passage

of others, probably by leaving some ions in its path, and by the increase in ionic mobility due to rise of temperature. Hence we reach an explanation of the phenomena of the electric arc, which is

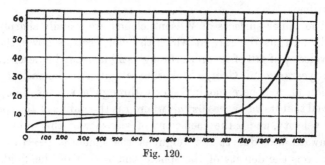

Fig. 120.

started by contact between the carbons, and persists as they are separated from each other.

In vacuum tubes (Figs. 110 and 111, p. 280), the high potential-gradient near the cathode causes the negative ions to start with very high velocities. J. J. Thomson and others have framed theories on the lines indicated to explain the complicated phenomena observed in such tubes.

**108.** We have stated that the electric charge on an ion of a gas is equal to that on a monovalent ion in a liquid electrolyte. We must now describe the experiments by means of which this result has been established.

Gaseous ions as condensa- tion nuclei.

The first series of experiments deal with the condensation of water-vapour on gaseous ions as nuclei. Long ago it was shown by Aitken that particles of dust facilitated the condensation of clouds, so that air would remain supersaturated when suddenly expanded adiabatically, if no dust were present. These results were extended by C. T. R. Wilson, who showed that, when the dust was removed by a few preliminary expansions, an expansion in volume from 1 to 1·38 was needed to produce a dense cloud. If the gas were ionized by X-rays, or by the radiation from radio-active substances, the cloud formed at an expansion of 1·25, while, by separating the ions by an electric field, it was shown that, although negative ions alone gave a cloud at an expansion of 1·28, positive ions needed an expansion of 1·31.

This greater efficiency of the negative ions as condensation nuclei may have a meteorological significance. Negative ions may be removed from the air by rain, leaving the air positively electrified.

The upper surfaces of Wilson's clouds are usually sharply defined, and thus the rate at which the cloud settles down, that is, the velocity of fall of its drops through still air, may be determined. Many years ago, Sir George Stokes investigated mathematically the friction on a sphere moving slowly through a viscous fluid, and showed that its value was $6\pi r\mu v$, where $r$ is the radius of the sphere, $v$ its velocity, and $\mu$ the coefficient of viscosity of the fluid. The downward force due to the weight of the sphere is $\frac{4}{3}\pi r^3 (\rho - \sigma) g$, where $\rho$ is the density of the sphere and $\sigma$ that of the fluid. In this case $\sigma$ is small compared with $\rho$ and may be neglected.

Now, as the drop falls, its velocity increases at first, and the frictional force increases with it. The weight keeps constant, and eventually the two forces will be equal and opposite. The sphere will then fall with a constant or limiting velocity $v_1$. We now have

$$6\pi r\mu v_1 = \frac{4}{3}\pi r^3 \rho g,$$

or

$$r^2 = \frac{9}{2}\frac{\mu v_1}{\rho g}.$$

Thus the radius of the rain-drops may be estimated by observing the rate of fall of the cloud.

By calculation from the known expansion used, and the total volume of the vessel, we may find the total mass and volume of the water condensed in the cloud. Dividing the total volume by the volume of each drop, we obtain the number of drops formed. If the ions be not too numerous, each ion acts as a nucleus of condensation, and the number of ions is equal to the number of drops produced.

Using Wilson's method of condensation, J. J. Thomson determined the number of ions per cubic centimetre of a gas. Their velocity under unit potential-gradient is known by the investigations described in § 105. Hence, by measuring the current $c$ through the gas when acted on by a known potential-gradient $E/l$, it was possible to calculate $e$ the ionic charge from the equation (p. 300)

$$c = ne(u + v)\frac{E}{l}.$$

The value of $e$, given by experiments made in 1901–2, is

$$e = 3\cdot4 \times 10^{-10} \text{ electrostatic units of electricity}$$
$$= 1\cdot13 \times 10^{-20} \text{ electromagnetic units.}$$

In Chapter X, page 232, we calculated the change on one hydrogen ion in liquid electrolytes, and found that the probable mean value was about $2\cdot1 \times 10^{-10}$. Within the limits of experimental error the charge on a gaseous ion is equal to the charge on a monovalent ion in a liquid electrolyte.

**109.** If an ionized gas be passed through a fine metallic tube,

Diffusion of ions.

it emerges in a non-conducting state. The ions diffuse to the sides of the tube, and either adhere to the walls, or give up their charges. By passing a stream of ionized gas through such tubes, measuring the saturation current through the gas, before and after the passage, and correcting for recombination on the way, Townsend determined the coefficients of diffusion of ions into different gases.

The number of ions which pass per unit time across any unit area at right angles to the axis of $x$ is proportional to the difference in concentration $n$ at two points on opposite sides of the area, and may be written as $D\,dn/dx$, where $D$ is defined as the coefficient of diffusion. We may regard the ions as moving parallel to the axis of $x$ with an average velocity $\dfrac{1}{n}D\,dn/dx$. The ions, being in the gaseous state, will exert a partial pressure $p$, which, at constant temperature, is proportional to the number of ions per unit volume. The average velocity may therefore be written as $\dfrac{1}{p}D\,dp/dx$. But $dp/dx$ is the force along $x$ acting on unit volume of the gas; thus the velocity with which the ions move under unit force is $D/p$. If the ions be exposed to an electric field of intensity $f$, the mechanical force on the ions in unit volume is $fne$, where $e$ is the ionic charge. Hence, if $u$ be the velocity of an ion in a field of unit intensity,

$$u = ne\,\frac{D}{p}.$$

Now $n/p$ is the same for all gases at the same temperature, and the ions are gaseous. Thus, if $N$ be the number of molecules in a cubic centimetre of air at the normal pressure $P$,

$$u = eD \frac{N}{P}.$$

From the known values of the ionic mobilities, we know $u$ in this equation, and Townsend's experiments give us $D$. The value of $P$, the atmospheric pressure, is about $10^6$ dynes per square centimetre, and thus $Ne$ may be found. Its value proves to be identical for all gases examined, air, oxygen, hydrogen, and carbon-dioxide, and the mean result is

$$Ne = 1 \cdot 24 \times 10^{10}.$$

Now, in liquid electrolytes, one electromagnetic unit of electricity, or $3 \times 10^{10}$ electrostatic units, liberates $1 \cdot 23$ cubic centimetres of hydrogen at $15°$ C. and a pressure of $760$ mm. of mercury. In solution, these $1 \cdot 23N$ molecules exist as $2 \cdot 46N$ ions, and, if $e'$ be the charge in electrostatic units on the ion of hydrogen in liquids,

$$2 \cdot 46 \, Ne' = 3 \times 10^{10}.$$

Thus                                $Ne' = 1 \cdot 22 \times 10^{10}.$

Therefore we may conclude that, in these experiments, the charge on a gaseous ion is identical with the charge on a monovalent ion in liquid electrolytes.

In another set of experiments (1908) by the action of an electric field, Townsend forced ions produced by Röntgen rays through a grating and then on to a plate surrounded by a guard-ring. The apparatus was so arranged that, if there had been no diffusion, all the ions would have reached the central plate. Owing to the action of diffusion, however, some of them spread outwards, and reached the guard-ring. By comparing with an electrometer the current collected by the ring with that collected by the plate, Townsend determined the effect of diffusion. In this case, positive and negative ions could be examined separately, one or the other reaching the ring and plate according to the direction of the field. It was found that, while the negative ions always carried the unit charge, the positive ions sometimes were associated with a double charge.

**110.** In every case examined by these methods, the charge on a gaseous ion seems to be the same as the charge on a monovalent ion in liquid electrolytes or double that unit. Now, as we saw above, the negative ions in highly exhausted tubes—the cathode rays—consist of negatively electrified particles, moving with velocities about one-tenth that of light, and possessing a mass/charge ratio equal to about the eighteen-hundredth part of that of the hydrogen ions in liquids. If we assume that, here also, as in all other cases, the charge is equal to that of the liquid hydrogen ion, it follows that the mass of the cathode ray particle is about the eighteen-hundredth part of that of the hydrogen atom.

It is most important that such a striking result should be confirmed in cases where both $e$ and $m/e$ can be measured for the same particles. Such a case has been found by Thomson in the ions due to the incidence of ultra-violet light on a metal placed in a highly exhausted vessel.

A zinc plate is illuminated with ultra-violet light, and placed opposite and parallel to a second metallic plate connected with an electrometer, the gas surrounding the plates being reduced to a very low pressure. An electric force is established between the two plates, and the negative ions, produced at the zinc plate, are by this force urged towards the second plate. If no other agency were at work, all the negative ions would reach the second plate, and transfer their charges to the electrometer. Now let us imagine that a magnetic force is applied at right angles to the electric force and parallel to the planes of the plates. The magnetic force will deflect the negative particles from their original straight course, and their paths become cycloids. They travel out from the zinc plate, curve round, and approach it again. If the second plate is placed near enough to the first to intercept this curved orbit, all the ions will still reach the plate connected with the electrometer, and the rate at which that instrument gains negative electricity will not be affected by the presence of the magnetic field. If, however, the electrometer plate be moved away from the zinc plate till it lies beyond the path of the ions, it will receive none of them, and the establishment of the magnetic force should stop completely the supply of negative electricity to the electrometer. If $X$ be the

*The nature of gaseous ions.*

electric force and $H$ the magnetic force, theory shows that no ions should cross the space between the plates if the distance between them exceeds $2Xm/eH^2$, while below that distance the addition of the magnetic force $H$ should produce no effect on the rate of gain of negative charge by the electrometer.

The experiments which Thomson carried out by this method showed that no such sudden change could be produced. As the distance was diminished, or the magnetic field increased, at first the effect of putting on or taking off the magnetic force was small. Then a stage was reached at which a considerable effect was produced; while finally, in a third stage, the magnetic force cut off almost all the ions from the electrometer plate. This somewhat gradual change is explained if we suppose that the negative ions are not all formed at the surface of the zinc plate, but that, as the primary ions there produced move forward under the action of the electric force, they produce new ions by their collisions with the molecules of the gas. The ions are thus formed, not exclusively at the surface of the plate, but throughout a thin layer of gas near the plate.

These considerations indicate that, in the experiments we are now describing, the limit of the second stage, in which some but not all of the negative ions are stopped by the magnetic field, gives the distance at which those ions coming from the surface of the zinc plate just fail to get across the space between the plates. The expression given above then leads directly to a value for $e/m$, the ratio of the ionic charge to the ionic mass. Thomson found as the result $7\cdot3 \times 10^6$, a number which agrees extremely well with that deduced for cathode rays, namely, $7\cdot7 \times 10^6$.

With the negative ions produced by the incidence of ultra-violet light on a zinc plate, it is easy to repeat C. T. R. Wilson's experiments on the formation of clouds round ions as nuclei, and thus to determine the value of $e$, the electric charge associated with the same ions for which $e/m$ has been obtained. The result shows that, as always, the charge is the same as the charge on a monovalent ion in liquid electrolytes; and therefore for the ions due to ultra-violet light, as for the cathode ray particles, we may take the mass to be about the eighteen-hundredth part of the mass of the hydrogen atom. The result has been confirmed by Lenard, who used a somewhat different type of apparatus.

In all these experiments, then, we are compelled to admit the existence of particles smaller than the smallest of the known chemical atoms. These particles, to which Thomson has given Newton's old name of corpuscles, are associated with a quantity of negative electricity, which plays the part of a true natural unit Positive ions, while carrying an equal charge of the opposite kind of electricity, have never been found with these small masses; they seem always to be structures of at least atomic dimensions. We may provisionally explain the facts by the hypothesis that the corpuscle itself represents a unit charge of negative electricity. An atom of ordinary matter with one corpuscle beyond its normal number is an atom negatively electrified; an atom with a corpuscle detached from it is an atom positively electrified.

Isolated corpuscles are found in cathode rays, in the radiation from radio-active substances, and in gases at low pressures and high temperatures. As the temperature falls or the pressure rises, the negative ions approach the positive ions in size, and at ordinary pressures appear, by their coefficients of diffusion, to be at least of atomic size. The corpuscle must then be imagined to attach to itself one, or perhaps more, neutral molecules, and this complex structure to act as the negative ion.

We are now in a position to estimate the importance of the experiments which have shown that the mass of the corpuscle is independent both of the nature of the gas in which it is found, and also of the material of the electrode used in producing it. Not only must we conceive atoms to contain these more minute particles, but it is necessary to suppose that in all atoms, whatever be their nature, these particles are similar. The dream of an ultimate particle, common to all kinds of matter, has thus at length come true.

**111.** A flame, or the hot gases therefrom, will discharge an electrified body, and the air near an incandescent solid is found to be ionized. Elster and Geitel, O. W. Richardson, H. A. Wilson and others have shown that, as a platinum wire is heated gradually, it begins to emit positive ions at a temperature corresponding to a low red heat. The investigation of the influence of a magnetic force shows that these ions vary in size, some probably being molecules of the gas,

*Ions from incandescent bodies.*

and others molecules of the metal or even dust disintegrated from its surface. As the platinum is heated still further, negative ions also come off, ultimately in large excess. In *vacuo* the negative leak from platinum and carbon filaments is very large—from carbon it may even amount to as much as an ampere of current from each square centimetre of surface. The negative ions are then of sub-atomic dimensions, and are identical with the corpuscles obtained by other means.

The emission of corpuscles at high temperatures is not confined to solids. Thomson finds that sodium vapour also gives off a large supply, and the effect seems to be common to all kinds of matter at a white heat. Carbon is particularly efficacious, perhaps because it can be raised to a higher temperature than can metals. It is easy to demonstrate the existence of a measurable current from one limb of the carbon filament of an ordinary incandescent lamp to an insulated plate placed between the limbs.

Owing to the emission of corpuscles by an incandescent wire or carbon filament along which a current flows, the effective current-carrying area of the wire is increased. In *vacuo* a considerable fraction of the current might pass through the space surrounding the wire, which must become filled with corpuscles. Although in gases at ordinary pressures the emission of corpuscles is less copious, still, ionization will occur to an appreciable extent just round the wire, and a part, though perhaps a small part, of the current will pass along outside the substance of the wire.

The phenomena we are now considering must have an important bearing on cosmical processes. The photosphere of the sun contains large quantities of glowing carbon, and this carbon will emit corpuscles until the resultant positive charge left on the sun exerts an electrostatic force great enough to prevent further emission. In this way a condition of equilibrium would be reached. Any local elevation of temperature would then cause a stream of corpuscles to leave the sun and pass into the surrounding space. When corpuscles pass through a gas with high velocity, they make it luminous, and Arrhenius has explained many of the periodic peculiarities of the Aurora Borealis by the supposition that cor-puscles from the sun, due either to incandescence or to some other cause, stream through the upper regions of the earth's atmosphere.

**112.** The phenomena of electrolytic conduction through liquids,
<br>The corpuscular theory of conduction. and of non-electrolytic conduction through metallic substances, must now be interpreted in terms of this corpuscular theory. The chemical decomposition of electrolytic solutions, which we have described in Chapter x, indicates that an electric transfer through such liquids involves a movement of the chemical constituents of the substance decomposed. In fact, as we have seen, that movement has been demonstrated experimentally, and the passage of the ions rendered visible. We must suppose, then, that the corpuscle forming the effective negative essence of the anion in liquid electrolytes is attached to an atom of matter. This atom may be associated with other atoms or molecules to form a complex ion, and almost certainly carries an envelope of solvent molecules with it, but the point is that the isolated corpuscle cannot slip from one atom to another, and thus carry an electric current through the liquid; the corpuscle cannot move without a corresponding movement of matter—of matter, that is, in its atomic or molecular sense.

Here again the motion of the positive ion involves the simultaneous passage of a particle of matter of at least atomic dimensions. The positive ion consists of an atom of the electrolyte with one of its corpuscles missing. A unit of negative electricty has been removed from it, that is, it is left with a positive charge. Electricity, on this hypothesis, is one, not two; the two so-called opposite electricities being perhaps an excess or defect of a single thing. Thus the nomenclature of the old "one fluid theory" is still appropriate. It is true that the meaning of the conventional names positive and negative should be reversed. But the signs conventionally attached to electric charges are a mere matter of historical accident. It is the old negative electricity which, from our present point of view, is the real corpuscle, and a positive charge means a deficiency in the proper number of the electrical units.

In metals, an electric current flows without chemical change in the substance of the conductor, so that, in this case, we must imagine the corpuscles to be freely mobile. They pass from atom to atom, and thus carry the current when an electromotive force acts. In the presence or absence of such a force, they may be regarded as existing within the metal in a state resembling in many ways the state of a gas in a closed vessel. Estimates have been made of the number of corpuscles present in a given volume; of the velocity with which they move under an electric force; and of their mean free path within the metal, that is, of the average distance a corpuscle moves between its collisions with other corpuscles. As we have seen, when the metal is heated, the corpuscles begin to leave it, and stream away into the surrounding space. At any constant temperature, equilibrium is set up between the corpuscles leaving the metal owing to the effect of temperature and those drawn back again by the residual positive charge on the metal. We may look on the system as analogous to a liquid in equilibrium with its own vapour.

In Chapters III and IX we saw that it was necessary clearly to distinguish the electric current and the heating effect of the current from the flow of the energy by which the current was maintained. The energy passes through the surrounding medium, through the luminiferous aether. The current is merely the line along which the energy of the aether can be dissipated as heat.

A conducting wire must be regarded as a channel along which the free ends of a line or tube of electric force can move, and, when the poles of a battery are connected by means of a wire, the tubes of force in the surrounding air run their opposite ends on to the wire, pull those ends towards each other, and shut up. Other tubes are then pushed into the wire by their mutual transverse pressure, and are obliterated in turn. The tubes of force in the dielectric field are thus inclined to disappear, and the state of aethereal strain in that field tends to be relieved. Simultaneously, however, the battery endeavours to reassert the original distribution of tubes, and once more to set up the strain. In this way new tubes are constantly forming between the terminals of the battery, and are as constantly pushed into the connecting wire, where they vanish. When the connection is metallic, it is only the negative ends of the tubes, attached to the corpuscles, that move, the positive ends remain at rest. If, on the other hand, part of the circuit be composed of an electrolyte, in that part the positive ends of the tubes, being attached to mobile ions, move also.

The ionic theory of electrolysis gave a clear idea of the mechanism by which the slipping of the ends of the tubes of force occurred in conducting liquids, and the corpuscular hypothesis gives us an equally vivid insight into the nature of the process within metallic circuits. The tubes, anchored by their ends to an ion in electrolytes or to a corpuscle in metals, drag their anchors. It is the slip of the anchors that constitutes the current, and the heat developed by the passage of the current is to be explained by the frictional resistance to the drag of the anchor, or to some other means of dissipating energy, such as inter-corpuscular radiation, not yet fully understood.

## REFERENCE.

*Conduction of Electricity through Gases*; by Sir J. J. Thomson.

# CHAPTER XII

## RADIO-ACTIVITY

Radio-active substances. Classification of the radiations. Radio-active change. Radio-active emanations. Induced or excited activity. The energy of radio-active change. Theories of radio-activity. Life of radium. Pedigree of the radium family. Radio-activity of ordinary materials. Electricity and matter.

**113.** THE discovery of the phenomena of radio-activity arose Radio-active indirectly from the study of Röntgen rays, and the substances. investigation of the new phenomena was rendered possible by the knowledge which was acquired previously and simultaneously of the conduction of electricity through gases.

Röntgen rays produce fluorescent effects on suitable screens; and it was natural to examine phosphorescent and fluorescent substances, to determine if they were the source of similar radiation. For some time no definite results were obtained; but, in the year 1896, M. Henri Becquerel discovered that compounds of the metal uranium, whether phosphorescent or not, affected a photographic plate through an opaque covering of black paper, even when no light had fallen upon the compounds previously.

It was soon found that, like Röntgen rays, the rays from uranium produced electric conductivity in air and other gases through which they passed. Compounds of thorium, too, were found to possess similar properties. In the year 1900, M. and Mme Curie made a systematic search for these effects in a great number of chemical elements and compounds, and in many natural minerals. They found that several minerals containing uranium were more active than that metal itself. Pitch-blende, for instance, a substance consisting chiefly of an oxide of uranium, but containing also traces of many other metals, was especially active. When

obtained from Cornwall its activity was about equal to that of the
same weight of uranium, but samples from the Austrian mines were
found to be three or four times as effective. The presence of some
more active constituent was thus suggested. To examine this point,
the various components of pitch-blende were separated chemically
from each other and their radio-activities determined. In this way,
three different substances, radium, polonium, and actinium, all
previously unknown, were soon isolated by different observers. Of
these three the most active was the well-known radium, discovered
by M. and Mme Curie, working with M. Bémont.

Radium is obtained from pitch-blende, in company with the
metal barium; and the two seemed at first to be connected chemically so intimately that the new substance was for a time called
"active barium." However, a slight difference in the solubilities of
some of their salts allows them to be separated gradually by a process of repeated fractionization, the radium chloride and bromide
crystallising out more readily than the corresponding compounds
of barium.

These processes of chemical separation are remarkable for their
use of the new property of radio-activity as a sole guide in the
operations. After each reaction the activities of both the product
and the residue were determined. It was thus settled whether the
reaction just tried was effective, and in which of the substances
separated by the reaction the property of radio-activity had been
concentrated.

The quantity of radium present in pitch-blende is extremely
small, many tons of the mineral yielding, after long and tedious
work, only a small fraction of a gramme of an impure salt of radium.

An interesting point in these investigations is the extreme
sensitiveness of the property of radio-activity as a test for the
presence of those substances which possess it. A delicate electroscope will show easily a leak of electricity with a substance having
an activity of about the one-thousandth part of that possessed by
uranium. The activity of pure radium has been estimated as about
one million times that of uranium; and such radium is a definite
well-marked chemical element, like other elements, forming salts
and other chemical compounds, and giving strong bright lines when
heated and examined with a spectroscope. Spectrum analysis has

hitherto been the most delicate means at our disposal for detecting
the presence of the chemical elements; but in the preparation of
radium from pitch-blende its spectrum only began to appear when,
in the prolonged process of fractionization, the product had reached
an activity of about fifty times that of uranium. Thus it appears
from these figures that the electroscopic method of detecting radio-
active matter is many thousand times more sensitive than the most
refined methods of spectrum analysis.

**114.** In the year 1899 Professor Rutherford, then of Montreal,
and now Sir Ernest Rutherford of Cambridge, discovered that the radiation from uranium consists of
two distinct parts. One part was unable to pass
through more than about the fiftieth of a millimetre of aluminium
foil, while the other part would pass through about half a millimetre
before its intensity was reduced by one-half. The first named, or
α rays, produce the most marked electric effects, while the more
penetrating, or β rays, are those which affect a photographic plate
through opaque screens. At a later date a third type of still more
penetrating radiation, known as γ rays, was detected. These rays can
traverse plates of lead a centimetre thick, and still produce photographs and discharge electroscopes. In proportion to its general
activity, radium evolves all three types of radiation much more
freely than uranium, and is best employed for their investigation.

The moderately penetrating, or β rays, can be deflected easily
by a magnet; and Becquerel, who deflected them by an electric
field as well, conclusively proved that they were projected particles
charged with electricity. M. and Mme Curie had shown previously
by direct experiment the existence of a negative charge associated
with these rays. Owing to their ionizing action, it is impossible to
demonstrate that a body surrounded by air gains a charge when
exposed to the rays. Such a charge would leak away as fast as it
was acquired. But, by working in a very good vacuum, or by
surrounding the body with a solid dielectric such as paraffin, the
acquisition of a negative charge can be demonstrated by means of an
electrometer. Further investigation showed that the β rays behave
in all respects like cathode rays, although they possess velocities
greater than those of any cathode rays hitherto examined, velocities

*Classification
of the
radiations.*

which have different values varying from 60 to 95 per cent. of the velocity of light. The $\beta$ rays, then, are negative corpuscles.

Magnetic and electric fields which are strong enough to deflect considerably the $\beta$ rays produce no effect on the easily absorbed $\alpha$ rays. Although R. J. Strutt, now Lord Rayleigh, in the year 1900, had suggested that the $\alpha$ rays were positively charged particles, of mass greater than that of the particles which constitute the negative $\beta$ rays, it was not till some time afterwards that their magnetic and electric deviations were demonstrated experimentally, and shown to be in the direction opposite to that observed with $\beta$ rays. Rutherford's experiments in 1906 gave for the ratio $e/m$ of the charge to the mass for the $\alpha$ particles a value of $5 \cdot 1 \times 10^3$. The value of $e/m$ for the hydrogen ion in liquid electrolytes is about $10^4$, since there is evidence given below to show that the $\alpha$ particles consist of helium, it follows that they are helium atoms (atomic weight 4) carrying double the univalent ionic charge. Their velocity is about one-tenth of that of light.

The very penetrating or $\gamma$ rays have never been deflected, and there is much evidence to show that they are different in kind from the other types, and, like the X-rays discovered by Röntgen, consist of waves of the same nature as of light. Further, it seems that like characteristic X-rays they consist of a series of monochromatic constituents characteristic of the emitting body.

Each type of radiation is absorbed by matter of all kinds in direct proportion to the density. The $\alpha$ rays, which can penetrate only some hundredth of a millimetre of aluminium, are absorbed by passage through a few millimetres of air. About 95 per cent. of the total ionization produced by a layer of radium or uranium is due to the $\alpha$ rays. Hence the ionization is much more intense in a layer a few millimetres thick lying next the radio-active material. This result is to be borne in mind in applying simplified theories, such as those given on pp. 296, 300 to these cases. The potential-gradient, too, between two electrodes, on one of which is spread a layer of uranium or radium, will not be uniform, being less where the ionization, and consequently the conductivity, is greater.

In the year 1905, a new type of radiation was discovered by Sir Joseph Thomson. When radium or polonium is placed in a vessel exhausted to the highest possible degree by the aid of charcoal cooled with liquid air, the substance is found to emit very slowly moving particles, which impart a negative charge to a gold-leaf suspended within the vessel. These negative particles are absorbed by a much thinner layer of air than the α rays, and, except in a good vacuum, are absorbed before they can be detected. It is not yet certain whether or not these slow-moving particles are an invariable accompaniment of α radiation.

It is remarkable that the radio-activity of a substance as measured by the emission of α, β, and γ rays is not affected by changes of temperature. Heating to redness, or cooling in liquid air, equally seem to be without effect. Again, the radio-activity of any element appears to be independent of the compound in which that element is contained. In thin layers, when absorption by the substance itself is not involved, the activity of a given mass of radium is the same when present as chloride or as nitrate, while uranium the metal has the same activity when combined as uranium nitrate.

The property of independence of temperature seems to forbid the reference of the energy of radio-activity to any ordinary chemical changes, while the identity of the effect for an element in different compounds indicates that radio-activity is a characteristic property of the chemical atom.

**115.** The experiments just mentioned suggest that radio-activity is an atomic phenomenon. But in the year 1900, Sir William Crookes found that, if uranium were precipitated from solution by means of ammonium carbonate, and the precipitate were dissolved in an excess of the reagent, a small quantity of insoluble residue remained. This residue, to which Crookes gave the name of uranium-X, was found to be intensely active when examined photographically, while the re-dissolved uranium was photographically inert. Similar results were obtained by Becquerel, who found that, when put aside for a year, the active residue had lost its activity, while the inactive uranium had regained its original radiating properties.

In 1902, Rutherford and Soddy discovered a corresponding effect with thorium, which may be deprived of part of its activity

*Radio-active change.*

by precipitation with ammonia. The filtrate, when evaporated, yields a residue which is very radio-active. After a month's interval, however, this activity had disappeared, while that of the thorium had regained its initial value. The active residue, thorium-X, seems to be a distinct chemical substance, for it is only separated completely by ammonia. Other reagents which precipitate thorium do not separate it from thorium-X. On these grounds it was concluded that the X-compounds are separate bodies, which are produced continuously from the parent substances, and lose their activity with time.

The processes of loss and regain of activity were then studied in detail. We will take the case of uranium, which, in some ways, is simpler in its properties than thorium. Crookes and Becquerel measured the activity by the action of uranium on a covered photographic plate. The cover absorbs the $\alpha$ rays, and hence the $\beta$ rays alone are used in such experiments. If the electrical method be employed, all the types of rays may be studied. Rutherford and Soddy found that the non-separable activity of uranium was due to the emission of $\alpha$ rays alone, while the activity of the uranium-X was due solely to $\beta$ and $\gamma$ rays.

The rates at which the activity of the uranium-X decayed, and that of the separated uranium recovered, are shown in the curves of Figure 121. It will be noticed that the curves are complementary

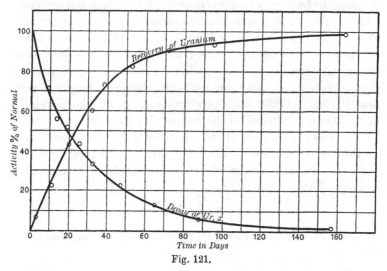

Fig. 121.

to each other; the activity regained by the uranium is, at any time, equal to that lost by the uranium-X. Again, we see that it is no more possible to change the total amount of radio-activity by chemical than by physical processes.

The activity of the uranium-X falls to half its initial value in about 22 days, and the residual half-activity again sinks to half in an equal period. This rate of change in a geometrical progression is typical of radio-active processes, the particular time-constant involved depending upon, and characterizing, the particular substance involved. Thus similar curves are obtained with thorium and thorium-X, but the time needed for the activity to become halved or doubled, in this case, is only 4 days. The determination of the rate of decay of radio-activity may be carried out easily by electrical measurements; it gives the most convenient and delicate test for the different radio-active substances, which may be distinguished from each other by this means.

In all such processes as those we are studying, the variable quantity changes by an amount per second proportional to the value of the variable at the beginning of the interval. Examples of this method of variation are given by the decrease in the pressure of the atmosphere as one ascends vertically from the surface of the earth; by the increase of a sum of money put out at compound interest; or by the decrease in the amount of a chemical compound which is dissociating, molecule by molecule, into simpler products. When a chemical change is brought about by the interaction of two or more molecules different laws hold. Thus the likeness between radio-active changes, which, as we have seen, are probably atomic phenomena, and mono-molecular chemical changes is significant. It suggests that the X-products are formed by the dissociation of the chemical "atoms" of the radio-active elements.

Let $I$ be the intensity of the activity of unit mass of uranium-X at the end of a time $t$. Then the rate of decay of activity is $-dI/dt$, and, by experiment, is proportional to $I$, the actual activity. Thus

$$-\frac{dI}{dt} = \lambda I,$$

where $\lambda$ is a constant expressing the fraction of the activity lost in unit time.

If we integrate this expression, taking $I_0$ as the initial activity, we get

$$\log_e \left(\frac{I}{I_0}\right) = - \lambda t,$$

which, by the nature of logarithms, may be written

$$\frac{I}{I_0} = e^{-\lambda t}.$$

On the theory which regards radio-activity as the accompaniment of the production of new substances, these equations denote the rates of formation and decay of the new radio-active material.

**116.** In the year 1900 Rutherford made another striking dis-
Radio-active    covery. The radiation from thorium was known to
emanations.    be very capricious, being affected especially by slight
currents of air passing over the surface of the active material.
Rutherford traced this effect to the emission of a substance which
behaved like a heavy gas having temporary radio-active properties.
This emanation, as it was named, is to be distinguished clearly from
the radiations previously described, which travel in straight lines
with high velocities. The emanation diffuses slowly through the
atmosphere, as would the vapour of a volatile liquid. It acts as an
independent source of straight line radiations, but suffers a decay
of activity with time, following the usual exponential law. Similar
emanations are evolved by radium and actinium, but not by uranium
or polonium.

The amount of emanation evolved by radio-active substances is
extremely small. A minute bubble has been obtained from some
decigrammes of radium bromide, but, in ordinary cases, the amount
is much too small to affect the pressure in an exhausted vessel, or
to be detected otherwise than by its property of radio-activity. It
is usually obtained mixed with a large quantity of air, and can be
transferred from one vessel to another with the air.

Its activity may be tested by introducing the air mixed with
the emanation into a brass cylinder (Fig. 122) along the axis of
which lies a brass rod fixed through an ebonite plug. The rod is
connected with an electrometer, and the cylinder with one terminal
of a battery. From the rate of deflection of the needle of the

electrometer, the current through the gas, and thus the ionization
due to the contained emanation, may be estimated.

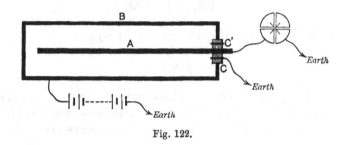

Fig. 122.

The radio-active emanations seem to be very inert chemically;
in this, they resemble gases of the argon group. They pass unchanged through acids or hot tubes, but are condensed at the
temperature of liquid air, evaporating again as the tube is warmed.

By measuring the rates of diffusion of the emanations into other
gases, their densities have been determined approximately, and
found to be about one hundred times that of hydrogen.

The rates at which the activities of the emanations decay
may be examined by storing a quantity of air mixed with
emanation, and, at regular intervals, placing an equal volume of it
in a testing cylinder such as that of Figure 122. If the ionization
be measured immediately, it will represent the activity of the
emanation alone, uncomplicated by the activity excited on solids
by contact with the emanation—an effect which we shall examine
presently.

The activity of the emanations is found to follow an exponential
law of decay, and the curves resemble those for the X-products
illustrated in Figure 121 on page 321. The radium emanation
decays to half value in about 3·7 days, and the thorium emanation
in about one minute.

Unlike the "straight line" radiations of the types $\alpha$, $\beta$, and $\gamma$,
the emanations are emitted much more freely from some com-
pounds of the radio-active element than from others, while the rate
of emission is largely dependent on physical conditions, such as the
temperature of the system. By a striking series of experiments,
however, Rutherford has traced these differences to variations in

the ease with which, after formation, the emanation escapes from
the generating substance.

Let us consider these results in more detail.  It is found, for
example, that while the emanation is given off very slowly from
dry and solid radium chloride, it is emitted freely from the same
salt in solution.  This allows the problem to be submitted to the
test of quantitative experiment.  The rate of decay of the radium
emanation is known; its activity falls to half value in 3·7 days.
Thus, the activity of the emanation stored in a solid radium salt
reaches a limit, when its rate of decay becomes equal to the constant rate at which the emanation is produced by the radium.  On
the hypotheses that the emanation is formed at the same rate in
the solid as in the solution, that it escapes from the solution as
fast as it is formed, and that it does not escape appreciably from
the solid, it is clearly possible to calculate the amount of emanation
that should be stored in the solid, as compared with the amount
produced and emitted by the solution in a given time.

The calculation shows that 463,000 times more should be stored
in the solid than is emitted by the solution in one second.  Now if,
as supposed, the emanation is stored in the solid, this large amount
will be liberated instantaneously when that solid is dissolved in
water.  Rutherford and Soddy measured this rush of emanation by
its effect on an electroscope, and found that it was 477,000 times
greater than the quantity afterwards developed by the solution in
one second: a remarkable confirmation of the deduction from the
several hypotheses given above.

The effect of raising the temperature is similar to that of solution.  When a solid radium compound is brought to a red heat, a
rush of emanation takes place, which makes the initial emanating
power some hundred thousand times greater than that of the cold
solid.  This high rate of emission, however, does not last; it, also,
is due to the rapid escape of stored material.

By experiments such as these, the emanating power of radio-
active elements has been brought into line with their other radio-
active properties, and has been shown to depend only on the mass
of the element present, whatever be the state of combination in
which that element exists, and whatever be the physical conditions
in which the process occurs.

**117.** If a rod, such as that within the cylinder of Figure 122

Active
deposits.
on page 324, be exposed to the emanation of radium
or thorium, it will itself acquire radio-active properties, especially if negatively electrified while in contact with the
emanation. When withdrawn from the vessel containing the emanation, and placed in a testing cylinder as shown in the figure, it is
found to ionize the gas.

Again, if a platinum wire which has become active by exposure
to thorium emanation be washed with nitric acid it is unaffected.
With sulphuric or hydrochloric acid, however, it loses nearly all
its activity, while the acid, when evaporated, gives a radio-active
residue. This result indicates that the activity on the wire is due
to the deposit of some new type of radio-active matter, which has
definite reactions with different chemical reagents. In the light
of the conclusions at which we have arrived above, it is natural to
interpret this new matter as a product of the disintegration of the
emanation from which it arises.

For long exposures to the emanation, the decay of the excited
activity on a solid surface follows an exponential law. For short
exposures, more complicated curves are obtained. Figure 123 shows
the rate of decay of the excited activity due to an exposure for a
few seconds to radium emanation. At first sight it bears no resemblance to the regular logarithmic curve in Figure 121 on page
321, which is the typical example of radio-active change. Rutherford has shown, however, that this radium curve may be explained
by the hypothesis that the radio-active matter undergoes three
successive changes, each of which conforms to the usual exponential
law. The first change, to which is chiefly due the rapid fall of the
curve near the beginning, is half completed in about 3 minutes ;
the second change involves no radio-activity, and is half completed
in 21 minutes ; while the third change is again radio-active, the
half-value being reached in 28 minutes.

If the rod be exposed for a few seconds only, the first kind of
deposited matter, which has been called by Rutherford Radium *A*,
alone is produced while the surface is in contact with the emanation. When the rod is removed, its activity decays rapidly, and
the matter to which the activity is due passes into a new kind,
Radium *B*, which, in its turn, gives rise to a third, Radium *C*.

This change, however, is not accompanied by radio-activity, or, at all events, by radio-activity measurable by ordinary methods. The third kind of matter again changes, and this time the process is

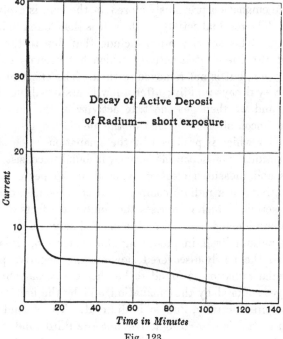

Fig. 123.

radio-active once more. The simultaneous process of these three kinds of change gives rise to the complicated phenomena represented by the curve of Figure 123.

If the rod be exposed to the emanation for some hours, a balance between these different processes is reached, and, when the rod is removed, the whole radio-activity decays in an exponential manner with the time.

Soon after appreciable quantities of radium were available for investigation, Giesel drew attention to the fact that a radium compound gradually increases in activity after formation, and only reaches a constant state after a month's interval. Similar phenomena have been observed by Curie and Dewar for the heat which such compounds emit (see p. 329).  These results are explained readily if we consider the properties of the emanation as elucidated by the experimental evidence that has now accumulated.

When a salt of radium is dissolved in water, and the solution boiled, the emanation previously stored in the salt is evolved and removed.  The residual activity of the salt is then found to be much diminished.  This activity must include that due to the radium itself, and also the excited activity, which has been developed by the emanation, but is not removed with it.  The effect of the excited activity decays rapidly; after a few hours it will nearly have vanished, and we then get the true activity of the pure radium salt alone, uncomplicated by that of the emanation, or by the excited activity which is produced by the emanation.

This residual, non-separable activity is found to consist entirely of $\alpha$ rays, and, measured electrically, is about 25 per cent. of the normal activity of a radium compound after a month's existence; a normal activity which comprises the combined effects of radium, of the radium emanation, and of the excited activity.

Thus, radium itself, in producing the emanation, gives rise to $\alpha$ rays and the newly-discovered slow-moving negative particles only; similar radiation is found to be the sole accompaniment of the changes suffered by the emanation, and by the first, transient kind of excited activity, due to Radium $A$.  The second kind of deposited matter is non-radio-active, but the third kind, decaying to half-value in 19·5 minutes, produces at least $\alpha$, $\beta$, and $\gamma$ rays.

Further changes have been detected.  Bodies exposed to the emanation of radium become radio-active, and a small part of this activity remains after the lapse of many months.  This persistent radio-activity has been traced to a substance, Radium $D$, with a time-constant of about 16·5 years, which although itself emitting no rays, passes successively into two active bodies, Radium $E$ and Radium $F$, with time-constants of 4·85 days and 136 days respectively.  Though these substances decay rapidly, they are continually reformed by Radium $D$, the forty years time-constant of which is too long to hold out much hope of quick improvement in the state of the walls, furniture and tools in laboratories where radium has been freely handled, and, in consequence, delicate measurements of gaseous ionization have become almost impossible.

**118.** In 1903 Curie and Laborde drew attention to the remarkable fact that compounds of radium constantly emit

The energy of
radio-active
change.

heat. They calculated from their experiments that
one gramme of pure radium would yield about 100
gramme-calories of heat per hour. More recent work shows that
one gramme of radium in equilibrium with its products gives 135
calories per hour. The rate at which this energy is emitted is
unchanged by exposing the radium salt to high temperatures, or
to the low temperature of liquid air, and certainly is not diminished
even at the temperature of liquid hydrogen.

The emission of heat was correlated by Rutherford with the
radio-activity. Radium freed from its stored emanation recovers
its radio-activity as measured electrically at the same rate as its
power of evolving heat, and the separated emanation shows variations in the heat developed corresponding with those observed in
its radio-activity. The electric effects of the radio-activity are
chiefly due to the $\alpha$ rays, and the heat effect also is chiefly dependent upon the emission of $\alpha$ rays. In the above-named total
of 135 calories per hour, only 5 calories are due to $\beta$ and 6 calories
to $\gamma$ radiation. The heat effect of the $\alpha$ and $\beta$ rays is clearly due
to the kinetic energy of the ejected particles.

**119.** The demonstration of the continual development of heat

Theories of
radio-activity.

by compounds of radium led to many attempts to explain the source of this apparently unfailing supply
of energy.

It was suggested that radio-active substances possessed the
power of absorbing the energy of some unknown radiation stream-
ing through space, and of giving it out again in the form of the
observed rays. Against this view it has been pointed out that
the radio-activity of radium is not affected by surrounding it with
thick screens of lead, which, it was argued, would presumably absorb some of the incident radiation. Moreover, the radio-activity
and emanating power have been shown to be the same when the
radium salt is concentrated, as when it is dissolved or spread in
thin films. Thus, parts of the radio-active substance itself, which,
by hypothesis, absorb this unknown radiation, do not appreciably
screen other parts from its effects.

Both these results, however, are possibly consistent with the hypothesis which supposes that the unknown radiation is absorbed only by radio-active substances. Thus lead would have no absorbing effect, or, at all events, none appreciable compared with that of radium. And the absence of screening by contiguous portions of radio-active matter is explicable if the intensity of the unknown radiation be great; a minute fraction of it would then alone be absorbed by the small quantities of radium which are available for experiment.

But the hypothesis is to be rejected on other grounds. It may be true that there is no evidence against it: at present there is certainly none for it. It introduces a new unknown: it is unnecessary, and nothing more damning can be said of a scientific theory.

As a result of their experiments on the emanations, and the excited activity produced thereby, Rutherford and Soddy explained the energy of radio-activity as due to the disintegration of the radio-active atoms. The evidence for this revolutionary hypothesis may be summarized as follows:

(1) Radio-activity is always found accompanied by chemical change; new substances are always detected after radio-activity has been going on—substances which have distinct physical and chemical properties, different from those of the pre-existent substances.

(2) The rates of development and decay of these new substances, as measured by their radio-activity, always follow an exponential law, whereby the rate of change is proportional to the amount of reacting substance present at any instant. This implies that the change is of the nature of a dissociation, in which individual particles alone are concerned. If the change were one of combination, or of double decomposition between two particles, the rate of change would depend on the square of the concentration, and another law would hold.

(3) Radio-activity is a characteristic property of the element—it does not depend on the state of combination, nor, in general, on the physical conditions, such as temperature, which affect ordinary chemical changes so decisively. The conformity with the

mono-molecular law of change indicates that single systems are dissociating; this further evidence points to the atoms as the systems involved.

(4) Since a gramme of radium emits 135 gramme-calories of heat per hour continually for many years at least, the total amount of energy liberated must greatly transcend the energy associated with any known chemical change. Hence we are led once more to look within the atom for the source of energy, and to suspect that the store of internal atomic energy is very great—an idea long ago suggested on other grounds.

Let us then, in terms of this new theory, restate the results which we have already described. All radio-active elements have very high atomic weight, the atom of radium, for instance, being about 226 times as heavy as that of hydrogen. Radio-active atoms are therefore very complex structures, and, on the theory we are considering, are capable of breaking down into simpler and lighter systems. The elements thorium and uranium for instance contain some few atoms which, at any moment, are disintegrating. As we have seen, the activity of the pure separated thorium or uranium consists of $\alpha$ radiation only. Thus, the essential process of the radio-activity of these substances consists in the emission of $\alpha$ rays, the disintegration of each atom resulting in the projection of one or more $\alpha$ particles with a velocity about one-twelfth that of light, while the residues break down into new and simpler atoms, which are themselves in a state of instability, and are known to us as thorium-X and uranium-X. The further transformation of these bodies is very rapid, their activity disappearing in a time to be measured in days.

As in the formation of the X-product, the essential process in the radio-activity which accompanies its disintegration consists in the ejection of the positively charged particle which we recognise as an $\alpha$ ray by its power of ionizing a gas through which it passes, and thus rendering that gas a conductor. The loss of this positive particle implies a change in the atomic residue, which now, in the case of uranium-X, seems to lose all radio-active properties, and therefore to pass out of reach of our present powers of observation.

In compounds of radium and thorium, however, we get the emanations as a step in the process of atomic dissociation. These bodies also are unstable, that is, radio-active. They emit new α rays, and produce the excited activity which is due to a deposit on the walls of the containing vessel. This deposit again breaks down, with the usual accompaniment of α radiation, and passes through the successive changes we have studied above, in the course of which β and γ rays appear.

In all these changes it seems to be the atomic nucleus which disintegrates, either with the emission of an α particle or the α-β particle associated with characteristic γ radiation.

The quantities of matter involved in any radio-active change are excessively minute, and no method at present known other than the electroscopic enables us to detect the final inactive products as they are formed. But, however, by the slow accumulation of material which must of necessity go on when a radio-active body is kept for a long time, the inactive products will be obtained eventually in amounts sufficient to be distinguished by the spectroscope or even by ordinary chemical analysis. In this connection we must give due weight to the fact that in radio-active minerals considerable quantities of the gas helium are occluded. Sir William Ramsay and Mr Soddy, by spectroscopic methods, detected the presence of helium in the gases evolved from a sample of radium, originally prepared from pitch-blende and kept as a solid for some months. The spectrum of helium was invisible when the emanation was first collected and examined, but it soon appeared and gradually increased in intensity with the lapse of time.

Similar results were obtained by Dewar and Curie, who, moreover, traced the disappearance of a minute volume of the emanation. This has been explained by the idea that the resulting helium, being projected in the atomic state as α rays with great velocity, penetrated the glass walls of the vessel and thus occupied no volume. The decrease in the volume of a minute quantity of emanation has also been observed by Ramsay and Soddy.

Finally, Rutherford and Royds compressed the emanation from 150 milligrams of radium bromide into a fine glass tube the walls of which were thin enough to allow the α particles to pass through them into an exhausted outer tube. After six days the gas collected from the outer tube showed the complete spectrum of helium.

**120.**  Radium itself being radio-active must, like its emanation,
Life of              be a transient substance.  Several methods have been
radium.              used by Rutherford to estimate the time required for
a given mass of radium to diminish by one-half, as its atoms disintegrate one after the other.

Since the mass and velocity of the $\alpha$ particles are known, their
kinetic energy can be calculated.  We can then estimate the total
amount of energy which one gramme of radium would emit by the
projection of $\alpha$ particles during its series of radio-active changes.  The
heat evolution gives the rate of liberation of this same energy.  Thus
Rutherford obtained the time 2000 years as the half-value period.

In 1905 he measured the quantity of positive electricity communicated to an insulated conductor by the $\alpha$ particles from a thin
film of radium bromide freed from the emanation and its products,
and concluded that the total current carried by the $\alpha$ particles was
at the rate of $4 \cdot 07 \times 10^{-9}$ ampere per gramme of the salt.  This
corresponds to an emission of $3 \cdot 1 \times 10^{10}$ $\alpha$ particles per second from
one gramme of pure radium.  If the disintegration of one atom of
radium to form one atom of emanation involves the projection of
one $\alpha$ particle, this same number gives the number of atoms of radium
disintegrating.  The total number of atoms in one gramme of radium
can be shown to be about $3 \cdot 6 \times 10^{21}$.  Hence the fractional number
exploding per second is $8 \cdot 6 \times 10^{-12}$ or $2 \cdot 7 \times 10^{-4}$ per year.  This result
gives the half-value period as 2600 years.

The number of $\alpha$ particles emitted by radium was also determined
by an ingenious method which enabled the effect of each single $\alpha$
projectile to be detected.  A gas at a pressure of a few millimetres
of mercury, when exposed to an electric field just weaker than that
required to produce a spark, is in a very sensitive state.  Any ions
formed in the gas are subject to strong electric forces, and acquire
velocities so great that they produce other ions by collision.  The
effect of the first ion is magnified in this way several hundred times.
Rutherford has calculated that one $\alpha$ particle produces 43,000 ions
before it is stopped.  In a strong electric field all these ions, and
all the secondary ions which each of these ions produces by collision,
will be swept to the electrodes before recombination can occur.
The electric effect of a single $\alpha$ projectile then becomes sufficient
to give a throw of the needle of an electrometer of some 25 milli-

metres on the scale. Using a very thin film of active matter at a distance from the opening into the testing vessel, the number of throws was reduced to 3 or 4 a minute. For radium, the results indicated that $3·4 \times 10^{10}$ $\alpha$ particles are projected per second from one gramme of radium and an equal number from each of its three $\alpha$ ray products. On these figures, the half-value period is 2300 years.

All these early values are too large. The calculation is now usually given thus. The number of $\alpha$ particles projected per second is $3·4 \times 10^{10}$. If $N$ be the number of atoms in one gram of radium $\lambda N = 3·4 \times 10^{10}$ where $\lambda$ is the disintegration constant. The atomic weight being 226, $N$ is $6·07 \times 10^{23}/226$.

Hence $\lambda = 1·26 \times 10^{-11}$, and $T = 1730$ years.

This has again been amended by later work and the most recent figure is 1590 years.

Another method by which the effects of individual $\alpha$ particles may be demonstrated is due to Mr C. T. R. Wilson. The clouds

Fig. 124.

produced by X-rays in air saturated with water vapour (see § 108) are caused also by α particles, and the track of each α particle may thus be made visible. Mr Shimizu has improved Mr Wilson's apparatus, and obtains repeated expansions in an air tight chamber by means of a reciprocating piston. Cloud tracks of the α and β rays can be seen or photographed at each expansion, which may be timed to occur from 50 to 200 times a minute. Fig. 124 shows α ray tracks, in some of which the sudden change of direction due to collision with an air molecule is visible.

**121.** Since radium is a substance short-lived compared with the age of the earth, either the small quantities which now exist must be the remnant of a store once vastly greater, or else radium must continually be formed from some parent element. If the total quantity of radium is constant, it must be formed at a rate equal to that at which it decays.

Pedigree of the radium family.

It has been shown by Strutt and by Boltwood that in the uranium minerals, from which radium is obtained, the content of radium is proportional to that of uranium. From uraninite containing 75 per cent. of uranium, to monazite with 0·30 per cent., the ratio of uranium to radium is constant within about one part in ten.

This relation suggests that we should look to uranium as the source of radium, though whether it is the immediate parent or a more remote ancestor remains a question for further inquiry. If it were the immediate parent, and the two elements in radio-active equilibrium in the minerals, radium must be produced in those minerals as fast as it decays. From this result it is easy to calculate the rate at which a given mass of uranium should produce radium, and to show that the amount developed by a kilogramme of a uranium salt in one month should give a quantity of emanation easily measurable by an electroscope. Experiments by Boltwood and others showed that the rate of production, if sensible at all, was much less than that indicated by the assumption of direct parentage.

Boltwood then showed (1907) that a new radio-active element chemically resembling thorium could be separated from uranium minerals by a series of chemical precipitations, and that, from this element, to which he gave the name of ionium, radium was produced at a definite and constant rate.

We therefore have the following pedigree of this family of radio-active elements.

| | Atomic number | Atomic weight | Time of half decay | Radio-activity |
|---|---|---|---|---|
| Uranium I ... ... | 92 | 238 | $5 \times 10^9$ years | $a$ |
| ↓ | | | | |
| Uranium $X_1$ ... ... | 90 | 234 | 23·5 days | $\beta, \gamma$ |
| ↓ | | | | |
| Uranium $X_2$ ... ... | 91 | 234 | 1·17 minutes | $\beta, \gamma$ |
| ↓ | | | | |
| Uranium II ... ... | 92 | 234 | $2 \times 10^6$ years(?) | $a$ |
| ↓ | | | | |
| Ionium ... ... | 90 | 230 | $2 \times 10^5$ years(?) | $a$ |
| ↓ | | | | |
| Radium ... ... | 88 | 226 | 1590 years | $a$ |
| ↓ | | | | |
| Radium Emanation... | 86 | 222 | 3·85 days | $a$ |
| ↓ | | | | |
| Radium A ... ... | 84 | 218 | 3·05 minutes | $a$ |
| ↓ | | | | |
| Radium B ... ... | 82 | 214 | 26·8 minutes | $\beta, \gamma$ |
| ↓ | | | | |
| Radium C ... ... | 83 | 214 | 19·5 minutes | $a, \beta, \gamma$ |
| ↓ | | | | |
| Radium D ... ... | 82 | 210 | 16·5 years | $\beta, \gamma$ |
| ↓ | | | | |
| Radium E ... ... | 83 | 210 | 4·85 days | $\beta, \gamma$ |
| ↓ | | | | |
| Radium F (Polonium) | 84 | 210 | 136 days | $a$ |
| ↓ | | | | |
| Lead ... ... ... | — | 206 | Inactive | |

Radium $F$ is a transient element, and must disintegrate into some other substance. In considering the problem that thus arises, we must not omit to notice that lead is a general constituent of uranium minerals. Five $a$ particles, each of atomic weight 4, taken from a radium atom would give a residue not differing much in mass from an atom of lead, of atomic weight 207. It is probable that lead is the ultimate product of this family of radio-active elements.

**122.** A few years ago radio-activity was unknown. Now, it is difficult or impossible to find substances which show no signs of the property. Traces of the emanation of radium or, at all events, of an emanation possessing an identical rate of decay, have been detected in the water from many deep wells, especially in the medicinal waters of Bath and

*Radio-activity of ordinary materials.*

Buxton; in many samples of earth and clay; and in the atmosphere, particularly in the air of caves and caverns.

A wire, negatively electrified and exposed to the air, acquires excited activity resembling in all respects that caused by exposure to radium emanation, and it is natural to conclude that the slight traces of conductivity found in gases in all circumstances are due to radiation from some radio-active material, near or far.

The ionization of a gas in an enclosed vessel might be referred to the presence of some slight trace of a radio-active impurity in the substance of the walls, or to a specific radio-activity of that substance itself. The solution of this problem is a task of much difficulty owing to the prevalence of traces of radium. Experiments by N. R. Campbell and others indicate that the metal potassium is quite appreciably, and some other substances are to a slight degree, radio-active independently of the presence of radium or other impurity.

**123.**  The success of Maxwell's electromagnetic theory of light
Electricity indicated that, as we have seen on page 209, the atoms
and matter. of material bodies, which, when incandescent, give
rise to light, must contain electric systems capable of vibration. The frequency of undulation of a wave of light is too great for us to refer the origin of the waves to atoms vibrating as wholes within the molecules, and we are thus driven to look within the atom for the vibrating particles. The complexity of the light emitted by many incandescent elements shows that the structure of the atom must be very complex also, and that the electric systems, which by their movements emit the electromagnetic waves of light, must be capable of many modes of vibration.

To explain these phenomena, a theory was framed, chiefly by Lorentz and Larmor, whereby the atom was imagined to be composed of a number of electric units, called electrons, in rapid orbital movement.  On this view, matter is a manifestation of electricity, and it was attempted to explain the further question thus raised by imagining an electron to be a centre of strain in the all-pervading aether, a place where, "to use a crude but effective image, the continuity of the medium has been broken and cemented together again without accurately fitting the parts."

If the waves of light take their origin from the acceleration of intra-atomic electrons, which, when in motion, must be equivalent to currents of electricity, a magnetic field should change the path of the electrons, and thus affect the nature of the radiation. By such reasoning the phenomenon of the doubling and tripling of spectral lines in a magnetic field was predicted before it was discovered experimentally by Zeeman.

When, by the experiments described in the last chapter, J. J. Thomson demonstrated the existence of negatively electrified corpuscles much smaller than the least of the so-called atoms, and pictured the atom as made up of such corpuscles, they were at once identified with the electrons, or isolated electric units, of Lorentz and Larmor.

Moreover, on the electric theory which pictures matter as made up of electrons in orbital motion, it was difficult to see why atoms should be invariably permanent and stable systems. Acceleration of the electrons would cause radiation and loss of energy, and systems stable while in rapid motion should sometimes break up as the motion became slower. Hence, when the phenomena of radio-activity forced the experimental physicist to face the possibility of atomic disintegration, a difficulty in the path of the electronic theory of the mathematician was removed. The occasional instability which he had foreseen as a consequence of his theory was found to be indicated as a more direct consequence of experiment.

Conceiving the atom as composed of a number of negative corpuscles or electrons, placed within a sphere of positive electrification in order to secure an inward force, Thomson deduced many of the observed periodic properties of the chemical elements with regard to their valency and to the nature of their spectra. Under the influence of their mutual repulsion, and of the central attraction of the hypothetical positive sphere, the corpuscles assume positions of equilibrium similar to those shown by a number of little vertical magnets, floated on water by corks, and drawn together by a large magnetic pole placed above them. As the number of magnets is increased, different patterns are formed, and patterns somewhat similar to each other will recur with stated numbers, thus recalling to mind the recurrence of similar chemical elements at regular intervals in Mendeleeff's periodic table, in which the elements are arranged in ascending order of their atomic weights.

But a modification of this picture of the atom, due to Rutherford and Bohr, is now generally accepted. The paths of the α particles from radio-active substances are usually straight, but sometimes, as we have seen, a sharp change in direction occurs. The forces exerted by the electrons on such a particle must be small, so that they cannot produce the change. But the phenomena are explained if it be assumed that an atom is a complex system with a positive charge concentrated in a minute nucleus, and negative electrons revolving round it in planetary orbits. Since the normal atom is electrically neutral, the positive charge on the nucleus must be equal in amount to that of the sum of the electrons in the atom, and, since the mass of the electrons is small compared with that of an atom, the mass of the atom must be nearly all concentrated in the nucleus.

If we combine these arguments with Moseley's work on atomic numbers (see p. 284) we get the modern theory of the atom. A hydrogen atom consists of a positive nucleus of unit mass and unit charge with a single negative electron, with a mass about 1/1800 of that of the nucleus, moving round it in a planetary orbit. Other atoms consist of a complex nucleus containing $N$ hydrogen nuclei or positive units bound together by a certain number of negative electrons, where $N$ is Moseley's atomic number, that is, the number of the place the element occupies in the periodic table in which hydrogen stands first. $N$ also represents the total number of electrons in the atom, and $Ne$ the total negative charge of all these electrons together, neutralized by an equal positive charge in the nucleus.

Since atoms can be ionized and given one, two, three or possibly four unit charges in accordance with their chemical valency, a small number of electrons are more easily affixed or detached than others and can be added to or subtracted from an atom with no complete change in its nature.

It used to be thought that any electromagnetic system subject to acceleration would emit radiation. The atomic electrons are accelerated towards the nucleus, since the direction of their movement is continually changing towards the nucleus as they swing round it in their orbits. But modern views regard radiation as discontinuous radiant energy being emitted and absorbed in definite units or quanta. If this be so, an atom in a state of steady, stable motion does not radiate. But, as in Thomson's model atom, more than one stable arrangement of electrons may be possible, and a sudden catastrophic change from one stable state to another may occur, as for instance by the passage of an electron from an outer to an inner ring, where its energy will be less. This loss of atomic energy appears outside the atom as a quantum of electromagnetic radiation, the frequency of which, to agree with other facts, must be supposed to be proportional to the amount of energy emitted.

Most radio-active processes are associated with the emission of an $\alpha$ particle, which must come from the atomic nucleus. Catastrophic changes, therefore, must occur in the nucleus itself. An $\alpha$ particle is a helium atom of mass 4, carrying 2 units of positive electricity. The residual nucleus will therefore be lighter and have a smaller charge. The external orbits of electrons will therefore have two electrons too many for neutrality, a rearrangement will take place and a new atom result.

Some radio-active transformations involve the emission of $\beta$ particles only. These too, by analogy, are supposed to come from the nucleus. In this case there will be no change in the mass of the atom, but, as the charge of the nucleus has been altered, the properties of the new atom will differ from those of the old one, and again a transmutation of elements will occur.

The conditions which bring about nuclear instability as we have seen are not affected by any known external agency. They seem to depend on chance happenings within the atom, and their frequency conforms to the well known laws of probability.

A moving electrified sphere carries with it the tubes of electric force represented in Figure 97 on page 210. Moving tubes of force imply a magnetic field proportional to their number and velocity, and hence to set such a sphere in motion requires the expenditure of the work necessary to establish the energy of the magnetic field. When once established, the momentum of the moving tubes, studied on page 215, will tend to keep them in motion; thus, owing to its electric charge, the effective inertia of the sphere is increased.

Let us apply this result to the case of a moving corpuscle. Theory shows that, for moderate speeds, the electric inertia outside a small sphere of radius $r$ surrounding the electrified particle is given by $2e^2/3r$, where $e$ is the charge on the particle. But, as the velocity of light is approached, this electric mass grows very rapidly. Now, in the $\beta$ rays of radium, we may examine negatively electrified particles moving with different velocities, some of which approach that of light. By measuring photographically the magnetic and electric deflections of these $\beta$ particles, Kaufmann has shown that the ratio $m/e$ of the mass to the charge of these particles increases as they near the velocity of light.

On the electron theory, matter is an electric manifestation, and the mass of a body must be explicable as electric inertia. On the assumption that the whole of the mass is electrical, Thomson has calculated the ratio of the mass of a corpuscle moving with different speeds to the mass of a slowly moving corpuscle, and compared these values with the results of Kaufmann's experiments.

| Velocity of Corpuscle in Centimetres per second | Ratio of Mass to the Mass of a slowly moving Corpuscle | |
|---|---|---|
| | Calculated | Observed |
| $2\cdot36 \times 10^{10}$ | 1·65 | 1·5 |
| $2\cdot48 \times 10^{10}$ | 1·83 | 1·66 |
| $2\cdot59 \times 10^{10}$ | 2·04 | 2·0 |
| $2\cdot72 \times 10^{10}$ | 2·43 | 2·42 |
| $2\cdot85 \times 10^{10}$ | 3·09 | 3·1 |

On the theory of relativity this increase of mass would happen on any view of matter, simply as the result of our methods of measurement. Nevertheless, since our modern atom of matter is an electric system, and electricity possesses inertia there is no need to look further for an explanation of mass—it is an electrical manifestation.

We have now explained, or explained away, the most fundamental dynamical property of matter, and still the mind is unsatisfied. It asks at once: What is the nature of this electric inertia into which mass has been resolved?

As we have seen, in § 73, it is possible to represent the inertia of a magnetic field as due to electric tubes of force in motion in the luminiferous aether. In this way electric inertia is in its turn 'explained" as "mechanical" inertia—of the hypothetical substance invented to enable our minds to form a rational picture of other physical phenomena.

No doubt, in a certain sense, simplification is thus attained. All natural phenomena are referred to the properties of the aether, and a complete theory of the aether would give us a complete account of the physical universe. Nevertheless the mystery is but changed; it is not dissipated. We may have explained matter in terms of aether; how are we to explain aether?

It is not without significance that a book on a branch of physical science closes most naturally with an unanswered question.

### REFERENCES.

*Radioactive Substances and their Radiations*; by Sir Ernest Rutherford.
*Electricity and Matter*; by Sir J. J. Thomson.
*The Recent Development of Physical Science*; by W. C. Dampier Whetham. Chapters VI and VII.
*Science and the Human Mind*; by W. C. Dampier Whetham and Catherine D. Whetham.

# INDEX